The

INCORPORATION

of

LIGHT

and
other problems

How the mis-definition and under-reporting

of light has doomed unification

by MILES MATHIS

edited by JOSEPH HYDE

*Incorporate def.3 :to give material form to :embody

Cover photo by Jastrow, Albani Collection

Apollo, Palazzo Nuovo, Capitoline Museums

Dedicated to Apollo
Driver of Charge

CONTENTS

Note

from the Author

I have been asked by my editor to make this book a bit more reader-friendly. My first book was all business, as you might say, and I spent very little time there putting things into perspective or summing up or, least of all, making my reader feel comfortable in the new universe I was building. This may be because I have intuitively felt from the beginning that my set of physical theories was both easier to understand and easier to stomach than current theories, and I thought this was obvious to all from the first word I uttered on the subject about a decade ago. Any sensible reader should be sighing a long series of reliefs upon discovering my writings, and I never felt the need to coddle him or her much beyond that. However, as a nod to the world outside my head, I have made some small attempt to ease the reader into this set of papers more gently, it may be, removing him somewhat less forcibly and with less acceleration from the jaws of the mainstream mythmakers. Some of you may thank me for this padding and some of you may curse me for it, but the cursers would be more efficient if they simply skipped ahead to the main course: no one is checking to ensure they read every word.

Since the upcoming preface supplies this overview and

cuddle, I will use the rest of this note to enlighten you on a more pedestrian matter: how the book is set up, and why. My first book was basically a best-of introduction to my papers, with some emphasis on my most important unified field discoveries. There I led by showing my reader that both Newton's and Coulomb's famous equations were already unified. This book will follow the same pattern, again being a best-of compilation, but with emphasis on a new unified field equation: the Lagrangian. Some have thought it would be better to divide the books by subject, and were I sure I were publishing a complete set I should agree. But, although I have been accused of over-confidence, I am not so sure of that. Who knows what the market will bear, or what the world will be tomorrow? Better, I feel, to get the most important papers out first, and worry about a complete set when there is a sure demand for it. There will be plenty of time for "complete works" in the future, and besides I have to leave something for future editors to do.

As it is, what we have here is about half a book worth of my charge field papers, with extensive bonus material after that. Since publishing my first book a year ago, I have published 50 new papers, and the bulk of these concern the charge field. My greatest discovery in the past year has been that the Lagrangian is a third unified field equation (UFE), and I have been able to completely unwind it, showing precisely how the terms work. I even show you how to go from my own UFE to the Lagrangian, in a few short steps of math.

Now a few words about the title of this book. This wording was chosen to represent the entire problem with light that is revealed in these papers. *Incorporation* literally means "giving a body", and up to now light has had no body in the

2

field. In mainstream theory, the photon has no mass and no radius, which means it has no real body. This should have been seen as a curious fact from the beginning, since it gives us an entity in physics that is non-physical. It turns out that this lack of real **incorporation** is what has kept the electromagnetic field from being unified into the gravitational field. The electromagnetic field is composed of photons, at the ground level, and an unincorporated photon simply cannot be unified. It has no real presence. Its presence is virtual only, and you cannot unify the virtual with the physical.

As the simplest example, consider the famous photo-electric effect, whereby electrons are knocked out of matter by photons. If photons have no mass or radius, how do they physically collide? How can an entity with no radius collide with anything? Only bodies collide, by the definition of "collide." To accept current physics, we have to throw out the definitions of words, both nouns and verbs.

Beyond that, I show in this book, in the chapter on Dark Matter, that the photon field or the electromagnetic spectrum has been vastly under-weighed. The mass equivalence of the spectrum has never been recognized, and this is precisely what has caused the 95% mass gap in the universe, not dark matter.

Finding how the charge field (photons) fit into the existing field equations has allowed me to unwind a large number of big problems very quickly. Almost everywhere I look I find confirmation of the charge field at the macro-level, and my readers have been very useful in helping me look. I have good data coming in almost daily that confirms my equations and theories, and I must admit that I can't keep up with it. I am more sure every day that I have pushed over the long-sought-after domino, since everything around me is falling, and the clicks are almost too much to bear sometimes. It has been difficult to pull myself away long enough to compile this second book, and I have had to ask my readers to save their emails until after this hits the presses. I can either publish five new papers a month or publish a book: I can't do both.

The good news is that the foundations of my theory are now so broad and so solid that others are going to be able to do this work for me. In many ways, and for a lot of readers, the stage has already been set. I don't have to advise them how to solve new problems, since they now have the simple tools to solve them before I get there.

Preface

It has been known for millennia that the Earth rests upon
the back of a giant turtle. Only in recent centuries
has this knowledge been added to. In 1794, in one of the
high valleys of the Himalayas, one of the wise was asked,
"Master, what does the turtle rest upon?" The Master
answered: "It is turtles all the way down, my son." But now
that scientists have finally succeeded in mapping the universe,
a turtle controversy has arisen. It turns out that level 7,484,912
is occupied not by a turtle, but by a man dressed as a turtle.
It is not known how this will affect our other equations.

You probably aren't used to having a book on science and math open with a joke. But a sense of humor is crucial to existing in a world where even our greatest accomplishments contain large elements of the absurd. Some contemporary thinkers are of the opinion that we are very near to a complete understanding of the universe. I am far from agreeing with them. We have made some wonderful discoveries and are due a small dose of pride, I suppose. But the things we don't know so overwhelm the things we do that any talk of a full understanding is just bombast. Worse, it is *hubris*. It may even be a scientific sacrilege, with real curses attached to it. When we become too secure in our knowledge, we stop questioning. Failure to question is the ultimate scientific failure. Answers quit coming precisely when they aren't sought, and they aren't sought precisely when they are (erroneously) thought to be in hand. We are like the dog who discovers how to use the little flap-door and now considers himself master of the house. He lies in front of the fire and congratulates himself for his cleverness. He would be better outside chasing rabbits.

In this book I propose solutions for several of the greatest errors currently existing in physics and mathematics. I do not propose to solve *all* the greatest errors, of course, or even to know what they are. I only present the ones that have become known to me in my years of research. Many may find my list surprising or even shocking, since I do not seem to choose problems that are commonly acknowledged to exist. Rather I choose problems that are believed to have been solved. This, I realize, can have the appearance of caprice or insolence, but I have simply gone where my nose leads me. I suspect that the whole history of science has moved in much the same way, so I will not apologize for seeing problems where I see them.

Lest I be dismissed as a crank before my first equation hits the page (and this sort of dismissal has become pandemic in the field), I rush to add that I am not a so-called classicist, bent on refuting Relativity and Quantum Mechanics simply because they disturb my sense of balance or my love of Newton. I attack Newton as well, long and—I like to think—shockingly. Beyond that, I am convinced of time dilation and length contraction and the necessity of transforms. I simply do not believe that Einstein provided the correct transforms. Likewise, I believe in the accuracy and usefulness of many of the equations of QED. But QED is still in large part a heuristic math posing as a theory. Even Feynman admitted this before he died, to the chagrin of most in the field. QED is not "the final solution" until it is fleshed out with a coherent theory. I believe, contra current wisdom, that QED *will* be provided with a coherent theory, one that makes sense even in the macro-world.

I am not a classicist, nor am I in any of the other dissenting groups that are opposed to the standard interpretation of Einstein. That is to say, I am not proposing supraluminal theories or any other theories that go beyond the math and theory of Einstein. I am not proposing any new particles, forces, fields, or maths. All the major chapters and findings in this book (and my previous one) deal with straightforward mathematical analysis of famous historical papers and theories. For the most part, this analysis is high-school level algebra applied to these papers. In critiquing the calculus, some rather subtle number theory is used, but no higher math at all. This means that this book is unlike anything you have read or heard of before. It is not allied to the *status quo*, but it is also not allied to any of the dissenting groups. It is completely outside the 20th century argument, since it cannot be said to be ultimately pro-Einstein or contra-Einstein, pro-Newton or contra-Newton. How you would classify me is therefore more a matter of your own allegiances than mine, since I have none.

This book differs from all the other critiques I have seen of current theory in that my arguments are not mainly philosophical or even theoretical. They are mathematical. I rerun the original equations in the original papers and show where the specific mathematical errors are. In this I believe I may be the first. I am certainly the only one currently attacking physics in the way I do. Yes, there are a lot of scientists in disagreement with the mainstream, but my agonistic method—pulling apart famous papers line by line —is not the fashionable way to do science now. It hasn't been the fashionable way to do science for over a century. No, it is a very different world intellectually than the world

Einstein entered when he began publishing with *Annalen der Physik* in 1901. The field of physics had not yet closed itself off from "amateurs." It was remembered then that Newton was a sort of amateur—a mostly self-taught mathematician and physicist—as were many of the greatest scientists and mathematicians of history. Einstein was a bit of an amateur himself, as the stories of his patent office imaginings confirm. The "university professional" was still a thing of the future. Forty years later amateurs still existed, though in fewer numbers. Karl Popper was resented maybe, but he was respected by most. Einstein himself understood the necessity of philosophy in the intellectual sciences, and he tied his theory early on to various epistemologies and metaphysics. He found it just as important to learn to speak of Kant and Hume as to learn the equations of Riemann. He was the last to do so.

The next two generations of physicists would lose all respect for the past. First Relativity and then Quantum Mechanics were seen to supersede all the theories of the past, and history became a clean slate. Richard Feynman could speak of philosophers with open disdain, and even Einstein was given only lip service. Einstein's "regression" into philosophy and his quarrel with the Copenhagen interpretation of QED made him a dinosaur in his own lifetime. TIME magazine may have voted him the most important person of the 20th century, but mainstream physicists considered him a befuddled old classicist by the 1940's.

My mathematical and mechanical critiques therefore arrives at a rather inauspicious time. They could not be less welcome. Physics as mechanics has been on a downhill slide since the 18th century, and it bottomed out some time in the 20th. In the new theories, mechanics is little more than a nuisance, and it is rarely even given a nod. Like many other things (think "art"), physics has been redefined as its opposite. Physics is no longer explaining things *physically*, it is developing mathematical and computer models that allow you to skirt the physics completely. Physics has been replaced by heuristics, and in many places by simple pettifogging.

Physics, despite seeming to be at a very creative time historically—due to the theoretical freedom that the top physicists would seem to have—is actually quite rigid and dogmatic. There are certain things you do and certain things you do not do. Superstring theory is prestigious. Looking at basic algebra is not. Looking into the distant future is progressive. Looking at old dusty papers is not. Tying esoteric theory to time travel and science fiction and Star Trek and the Dalai Lama is *au courant* and cool. Tinkering with ancient history is not. Stephen Hawking can claim that physics will be over in ten years, since ten years is still in the future (and apparently always will be, by some paradox), and not break any unstated laws. But a scientist who claims that Einstein or Newton or Feynman may have made a verifiable mathematical error is seen as monomaniacal and anti-social. Despite all that, I am confident that my math will speak for itself with those who have eyes to read. It is to be hoped that I have left very little room for argument in my equations. Metaphysics may allow for endless bickering, but algebra was invented to finalize the argument. Even the tensor calculus may allow for some movement: there are places to hide amongst the matrices. With algebra there is no shelter as large as a shrub to huddle beneath.

I began this series of books when I stumbled across the first great error many years ago, in reading Einstein's *Relativity*.

Although it soon became apparent that the error was both elementary and profound, I thought at the time that it was an isolated error. But my naiveté evaporated as I subsequently reread other important theoretical papers, and my awe of the past evaporated with it. What I came to realize, with rising disbelief (as well as some excitement), is that my faith—the faith of all scientists—in the basic theory and math of physics has been unfounded. It became apparent that the theory and math of many famous and influential papers, both classical and modern, had never been checked closely—or not closely enough for my taste at any rate. Buried in these papers were algebraic and geometric errors of the most basic kind. Suffocating beneath dense, often impenetrable theories and unnecessarily difficult equations of so-called higher math were errors that a high school student could understand, were he or she presented with them in a straightforward manner.

My goal became to do just that. To strip physics of its mystifying math, its unnecessary proliferation of variables and abstract concepts, its stilted language and dry jargon, and to speak in clear everyday sentences and simple equations. Einstein is famous for stating that a theorist should be able to explain his theory to an eighth grader, but he did not practice what he preached. Like his precursors, he could not explain his theory even to his peers. Relativity has remained uncorrected for a century not because it is flawless but because, as written, it has been impervious to understanding.

As incredible as it may seem that errors have remained uncorrected in Relativity for a century, that time period is actually quite small compared to other errors I will relate here. The errors of Newton have persisted untouched since he made them, traveling unnoticed beneath the noses of the greatest mathematicians in history. And the errors embedded in the calculus are older still. We have to go back to ancient Greece to find the theoretical underpinning of Newton's and Leibniz's calculi. This theoretical underpinning was often improved upon in the 2000 years between Archimedes and Cauchy—which makes it all the more amazing that it is false. Mathematicians spent two millennia refining an error. The calculus is true, but its theory is false. It does not work the way anyone has ever thought it does, or for the reason anyone has ever thought it does. It has nothing to do with infinitesimals or limits. But I am giving away the ending of a great story.

It was fortunate that I discovered early on the soft underbelly of modern math, for it allowed me the rare privilege of transcending it. I saw almost from the beginning that esoteric maths such as the tensor calculus had become obstructions to true understanding. If the tensor calculus could build its greatest structure on the false math of Relativity, then it must be an overrated tool. An architect who knows his job does not build a palace on a sand pit, and the mathematician is a fool who spends his college years diddling with a math better done on computers, when he doesn't understand algebra or geometry.

As a tonic to this chaos, I have tried at each point in my proofs to use the simplest math possible. This runs counter to current dogma, which tells us to impress each other with the most difficult math imaginable at all times. Simple math is considered neither sexy nor imposing. It also cannot be used as ballast, as misdirection, or as obfuscation. It is therefore not of much use to the modern theorist. Careers are advanced by advanced math; nothing is propelled by simple algebra, it is thought. Despite this, I have found that algebra is the first and most useful tool for unraveling the mathematical mystifications of the past. In the beginning, Special Relativity was proved by Einstein with algebra. The 1905 paper has only one line of calculus in the proofs of the transforms, which line is redundant padding. These transforms are exactly the same ones proved today with the tensor calculus. But the obvious tool to critique algebra is better algebra.

In correcting the foundation of the calculus I did not need calculus or any math evolved from it. I only needed basic number theory, which basic theory is now so elementary as to be forgotten. The modern mathematical method for solving any problem is to come at it from above, with more and more abstract math. My method is to come at it from below, questioning the fundamental postulates and often simple math that have been lost to view over time. As an example, the problem of gravity is being attacked now with superstring theory, which preens itself on its mathematical complexity and its theoretical density. But I have shown that gravity can be solved by unlocking simple algebraic relations among classical variables. There is very much in the existing theories of Einstein, Lagrange, Hamilton, Newton, Kepler, Galileo, and even Euclid that had never been resolved. Leaving these mysteries in the trash in order to concentrate on new mathematical paradoxes has been a grave error in judgment.

Descartes (who also missed seeing the fundamental error of the calculus, by the way) said in his *Meditations* that he had reached a point of absolute doubt. He felt he could rely on nothing around him. He must start over from the beginning, taking as true only what he could prove himself. Most philosophers now believe that Descartes was only using a convenient method of argumentation, one that did not seem so unique, or so egotistical, in the 17th century. But

I believe he was in earnest. I find his doubt highly plausible, even beyond its usefulness in critiquing the unsupported beliefs around him. As more and more of the pillars of my certitude fell, I too reached a point of near-infinite doubt. I found that I could no longer look at any theory or equation, no matter how self-evident it seemed, without checking the math from top to bottom. No more would I take any proof on faith, assuming, as an example, that a short series of equations by Richard Feynman must be correct, simply because I knew that he was famous for being a great mathematical physicist. I have since found absurdly simple mistakes everywhere I looked. In fact, it has been rare that I have checked anyone's math and found it correct. I have gone through textbooks, finding algebraic errors on nearly every page. The calculus is almost universally misused, even beyond its cardinal error in claiming to find instantaneous values. The newer maths, many of them offshoots of calculus, are likewise flawed in many fundamental ways, from set theory to topology to Cantor's theory of infinities.

I know that most will be shocked at my presumption, and the rest will question my credentials. But I can only answer that physics has never, in the whole history of science, had anything to do with credentials or false humility. It has to do only with truth. If my equations are faulty, then I am abashed. If my theory is incomplete, I am vulnerable. But no one should have to apologize for having the courage to question, or to present his findings. The overly socialized and pressurized milieu we live in, where intelligent and earnest people are dismissed for the flimsiest of reasons, or for no reason, and where most people are cowed into permanent silence, has more to answer for to history, or to the gods of physics and math, than I ever will for my boldness.

Chapter 1

11

Eleven Big Questions
you should have for the standard model

We get a constant line of bald propaganda from physics now, claiming a near-complete knowledge of the universe. This propaganda is not new: it has been building for over a century. Lord Kelvin claimed (around 1900) that there was nothing new to be learned in physics. Relativity and quantum mechanics shushed the Lord Kelvins for a few decades, but soon they were at it again. The Big Brag hit what one might call a new crescendo in 1988 with the release of Stephen Hawking's book *A Brief History of Time*. There Hawking claimed that we would achieve physical omniscience within a decade, and physics would be finished. Now, over two decades later, we are no nearer omniscience; we are only nearer a perfect *hubris*. As I showed in my analysis of a recent NASA video on Hulu.com, most science releases meant for public consumption still lead with this claim of near-omniscience. We are told that we are close to complete understanding of physics, and that we need only a couple of small pieces to complete the puzzle.

I have written my papers mainly to counter this very disgusting (and unscientific) attitude. Now I am writing this paper to compile some of the biggest cheats and fudges I have uncovered, so that you may have them all in one place to refer to whenever you come face to face with one of these promoters of the "near-perfect" knowledge of modern physics.

I say "standard model" in my title because although the standard model usually refers only to the status quo in particle physics, there is a standard model of physics in every sub-field. This standard model has ossified into dogma, into a set of beliefs one is not allowed to question. Anyone in academia who questions the standard model can expect to be set upon by the institutional jackals and marginalized into oblivion. He will find his papers refused for publication and his funding cut off. Those outside academia will simply be dismissed as cranks and crackpots.

As a fair debater myself, I will tell you the number one rule of debating according to the current handbook: always keep your opponent on the defensive. It is not a device I use, since I prefer the more subtle and less used method of actually knowing what I am talking about. But the mainstream preferentially follows the number one rule, since in most cases it is so effective. They know that with most audiences, facts and truth mean almost nothing. Everything is judged on form, and if your opponent can be made to look awkward, you will have won more points than could ever be won by being right. For this reason, you are taught always to ask questions; never to answer them. Yes, this is the technique of the mainstream in dealing with any resistance. Always attack. Always go for blood. If there is a threat, do not address the substance of it. Attack the person.

Therefore, if you get into a debate with anyone over physics (or anything else), I suggest you remember what your opponent is up to at all times. Do not allow him or her to put you on the defensive. If your opponent is part of the status quo in physics, he or she should be able to answer questions. He is claiming near-omniscience, not you, so he should be able to answer all questions with ease, like the god he claims to be. The standard model is the one making the money and getting the attention and taking all the jobs, so they are the ones that should be answering questions, not you. They are the ones getting all the magazine and journal articles, all the book publishing, and all the government funding, so they are the ones that should be answering questions, not you. You are on the fringe, an independent researcher, a person just trying to help for free, so it is no surprise if your ideas are incomplete. No one should find that out of the ordinary. But what IS extraordinary is that mainstream physics, which has been gathered and culled by thousands of geniuses over centuries, and is defended by all the top people now, is full of huge awful holes and embarrassing fudges. Even more extraordinary is that these self-styled geniuses and top-of-the-field people do not have the intellectual honesty or the scientific acumen to see these holes and fudges for what they are, and to want to correct them. Remember that always.

So let us look at the questions I have had for the mainstream. I have tried to answer their questions, even when these questions are clearly hostile and poorly chosen. But they don't ever address my questions, or anyone else's. They just dodge and misdirect. They do this because there is no possible answer to my best questions. My best questions are immediately fatal, and I like to think they can see that. These eleven questions are among the most embarrassing and fatal questions in my papers, and you will never see the mainstream address them. These are the questions that have been in the dark, are in the dark, and will remain in the dark, if the mainstream has anything to say about the matter.

1) In the case of a gravitational resonance, as in the resonance with Jupiter and Saturn, what causes the bodies to begin moving apart after the closest pass in the resonance? Gravity is stronger at closer distances, so what makes the resonance "turn"? [I address this question in my first book in a chapter on Laplace, and I will address it in this book in the chapter on the C-orbit asteroids (chapter 12).]

2) Roche limits are an outcome of gravity, so why don't the inner moons of Jupiter and Saturn obey gravitational laws? They not only go below the Roche limit, and avoid break-up despite having low densities, they also survive large impacts (as we see from large cratering). Finally, they accrete. How can bodies that should be dissolving accrete?

3) We are told that atmospheric muons are experiencing time dilation in order to reach sea level detection. But special relativity tells us that all objects in relative motion experience both time dilation and length contraction. The length contraction in SR is derived from the x or distance contraction, and they are proportional. Meaning, the whole x-dimension must be contracting, not just the "length" of the muon. Which means that a time-dilated particle must seem to be going a shorter distance than expected, not a longer distance. How can current theory ignore the length contraction? [See my chapter on muons in my first book].

4) The orbit is currently explained by only two motions: gravity and the velocity of the orbiter. But according to Kepler's and Newton's equations, which still stand, this velocity is the tangential velocity. It is not the orbital velocity, since the orbital velocity is the result of the two motions, not the cause of them. In other words, the orbital velocity curves, and it curves because it is composed of the centripetal acceleration. If the orbital velocity is the result of the two motions, it cannot be one of the two motions. According to Newton, the tangential velocity is the "innate motion" of the orbiter. But this innate motion cannot curve by itself. Given these two motions, why is the orbit stable? Current physicists just sum to show the stability, but summing hides the variations in the differentials. The problem is that if we study the differentials, we find the tangential velocity varying to create the stability. How can the "innate motion" of an orbiter vary? Are we to imagine that orbiting bodies are self-propelled, or that they can change their motions to suit summed orbits?

5) Perturbations are an important part of solar system mechanics. These perturbations often take the form of torques or tangential forces from one body to another. Given that neither Newton's nor Einstein's fields allow for forces at the tangent caused by the gravitational field, how do physicists justify these torques?

6) Symmetry breaking is a common tool of modern particle physics. Since symmetry breaking requires borrowing from the vacuum, how is this physically justified? What are the rules for borrowing? That is, why can particle physicists borrow from the vacuum in order to fill holes in electroweak theory, but I cannot borrow from the vacuum to fill all the holes in my theories? Is it something to do with institutional credits? Is Goldman Sachs involved in this borrowing?

7) After more than a century of silence, the standard model finally assigned the "mechanics" of charge to the messenger photon, a single virtual photon that can either tell quanta to move away or move nearer. What is the operation of this "telling"? It is some sort of code etched on the virtual face of the virtual photon? Is it a mysterious wave sent across intervening space, a wave that can be inverted at the will of the photon? Or is it a voice message? A Tweet perhaps?

8) Speaking of virtual particles: is there anything a virtual particle cannot do? Are there any rules of virtuality? For instance, if virtual particles can explain charge and color and borrowing from the vacuum, why can they not explain every other problem of modern physics? Where is the imaginary line drawn, and why draw it there? Once you begin cheating, why cheat halfway when you can cheat all the way?

9) If e=mc^2, and if the photon has energy, how can it be massless? How can an equation with the speed of light in it not apply to light? Sure, we can say that the photon has no rest mass, since it is never at rest, but how can we say it doesn't have moving mass? Don't energy and field equations, like charge equations, have to be fudged, in order to deny mass to the photon? Energy without mass contradicts both the classical equation and definition of energy (e=mv^2/2) and the relativistic equation and definition of energy (e=mc^2). Might this be why particle physics now hides out in a renormalized gauge math?

10) If gravity is now defined by curvature rather than by a centripetal force, what impels an object placed at rest in a field to begin moving? General Relativity supplies us with field differentials, which can explain why an object already moving in the field will move as it does. But field differentials, being math, cannot create a force. The math of GR represents motions, it cannot cause them. GR is also not

a field of potentials, since it requires a field of forces to create potentials. GR is not a field of forces, so the differentials cannot be interpreted as potentials. Einstein admitted that GR was the bypassing of Newton's inertial field. How can an object that is "feeling no forces" begin moving in such a field? In other words, Einstein inherited and extended the field of Newton, but he did not overwrite Newton's first law. If he had, we would not still be taught it in high school. Newton's first law is that an object at rest will remain at rest unless a force acts upon it. What force acts upon an object placed in Einstein's curved field? How does the object know that the field differential just below it is any different than the field differential it inhabits? It can't know, and therefore GR fails to explain motion from rest in a field.

11) The Moon is experiencing tides front and back caused by the Earth. Because the Moon is in synchronous orbit, these tides are always in the same place: they do not travel. All tides are caused by two mechanisms, we are told. They are caused by different levels in the gravity field, and they are caused by unequal centrifugal forces due to the orbital motion. The second effect is half the first, so it is 1/3rd the total: very significant, in other words. If the forward and backward points of the Moon are experiencing strong and constant tides, why are they not shearing strongly sideways? The farthest part of the Moon should shear in the reverse direction of the orbital motion, since there is nothing in the gravitational field to make it orbit faster than the center of the Moon. Just the opposite, in fact. If we assume all parts of the Moon have the same "innate motion", and if we are given that an object at a greater distance has a smaller acceleration from the field, then the farthest part of the Moon should be going slower than the center of the Moon. As it is, it travels faster than the center of the Moon for no

21

physical reason. The reverse applies to the forward part of the Moon, and it should shear in the direction of orbit. Why is this data so obviously negative? Among other hugely embarrassing data on the Moon is the negative tide on the front. The standard model of tides predicts equal tides front and back, but the Moon's crust is obliterated almost down to its mantle in front, showing an obvious negative tide. The standard model has no explanation for this, while I have a simple and mechanical explanation. The question is, how can piles of obvious data like this continue to be ignored, when there exist straightforward explanations for it?

Readers sometimes write to me, telling that if I would just do this or that, I would blow the roof off physics and win a Nobel Prize. As you see, I have already blown the roof off GR and QED and Newton so many times the molecules won't even cohere into shingles anymore. An honest person would just admit that and ask what's to be done. Instead, the mainstream simply refuses the see the holes I have pointed out. They pretend that I have not asked them a thousand important questions, and they begin scanning my papers for weak points. That is also a clear sign: a real scientist would scan any paper for its strong points, since those are the most useful to science. Instead, mainstream scientists scan any new ideas, especially those from outsiders, for their weakest points, ignoring the strong points on purpose. This immediately proves that the reading is hostile, and therefore unscientific.

They redirect always: they pretend that this is not about them, because they want it to be about me. Remember their mantra: never answer a direct question, or look directly at a problem; instead, attack the questioner personally and ask him an unbroken line of questions, so that he can never be on the offensive himself. They always say something like, "You are the one claiming to know something we don't, so you should have to prove it." But that is just misdirection. It is true, but it is true on a much smaller scale than my reply to them, which is, "We are both claiming to know something, but you are the one whose account has been accepted. Therefore, it is even more important that your account be tested than mine. Besides, I admit doubts about both your theories and my own, while you admit no doubt about anything. Since your doubt is 100 times less, your data should be 100 times more secure. But it isn't. You have just unloaded all your negative data into a dark pit, and refused to remember it exists. You claim that physics should be testable, but then dodge all tests except those that you create to confirm yourselves. You look away from huge piles of negative data, and get mad when it is pointed out. That isn't scientific. Science requires criticism, but you refuse to countenance any criticism, blacklisting anyone that doesn't immediately accept your proposals. Your whole method of teaching makes this clear, since it is a method of indoctrination and peer pressure, rather than an open method of free inquiry. You have been defining science as free inquiry for hundreds of years, but the amount of free inquiry that actually gets done in academia is now near zero. Free inquiry in a time of such partial knowledge would spawn great disagreement and debate, and the fact that we have so little of either is clear evidence against free inquiry. Therefore, don't be surprised when I take your hostile questions with an ill grace. I can see them for what they are: suppression of science."

What I have done is to dig deep into the closet and pull that bundle of negative data back out into the open. I have laid all the old problems out on the sidewalk, where passers-by can see them and study them. I have pinned all the old data on the trees in the front yard, where it can air in the wind once more. For this reason, I hardly need new data or experiments of my own. This old data can be used by either the standard model or by me, and since they have no use for it, I am free to use it myself. This was a good move on my part, since this old data has turned out to be my best friend. As in a zero-sum game, every plus for me scores a minus for them, which is a change of two in the game. While they have been jacking themselves off with string theories and backward causality and virtual particles and symmetry breaking, throwing a series of airballs, I have been scoring at least two points with every paper I write. I passed GO some time ago, and they are still rotting in jail or in no parking, looking for the community chest. They no longer even have the gumption to realize that I own all four railroads, that the wheels are off their car, the shoe is off the foot, and the little silver hat is headless.

So print out this list and sew it into your peacoat, like Thoreau did with Carlyle's *Sartor Resartus*. And when some mainstream stuffed shirt starts calling you a crank for not bowing down before him and his false gods, ask him these simple questions. Do not let him dodge them, and do not let him re-direct the argument into some slur upon your alma mater or your IQ. Badger him, browbeat him, and always seek the higher ground where you can look down upon him. If you don't, he will do it to you. And if you ever feel the least bit unsure of yourself, return to the question he seems to like the least. Roll it up into a sharp point and metaphorically try to insert it into his ear.

Chapter 2

Death

by

Mathematics

The state of learning now is like Scylla of the old fable,
who had the head and face of a virgin,
but a womb hung round by barking monsters,
from which she could not be delivered. —*Francis Bacon*

In the 20th century, physics underwent a transformation. No one would deny that. But normally the transformation is credited to Relativity and Quantum Mechanics. And normally the transformation is seen as a great advance. In this paper I will argue the opposite. The transformation was due more to a transformation in mathematics, and that transformation has been almost wholly deleterious.

This transformation due to mathematics began in the 19th century, but it did not engulf physics until the 20th century. In the 19th century the stage was set: we had several abstract mathematical fields that reached "fruition", including a math based on action variables and principles, a math based on curved space, a math based on matrices, a math based on tensors, a math based on i, and a math based on infinities.

As I have shown, 19th century mathematics inherited many unsolved problems from the past, including problems from Euclid and Newton. It made no progress in solving these problems because it did not recognize them as problems. It had already given up on foundational questions as "metaphysics", and it preferred instead to create more and more abstract systems. The more abstract the mathematical system became, the more successful it could be in avoiding foundational questions.

The clearest example of this is the field of applied mathematics based on action variables. For the last hundred years we have heard an ever-increasing level of praise of action variables, culminating in the propaganda of Feynman. But action variables are just an abstraction of Newtonian variables. By abstraction, I mean that they do not add clarity, they cloak disclarity. Newtonian variables were never very rigorously defined, but action variables are very good at hiding Newtonian variables. Action variables do not replace Newtonian variables, as some appear to think. Action variables *contain* Newtonian variables. Action variables restate Newtonian variables in what is considered to be a more efficient form. But action variables are utterly dependent on Newtonian variables. If it were discovered that Newtonian variables were false, action variables would be, too, by definition. The action concept developed directly out of Newtonian mechanics, and action assumes the absolute validity of Newtonian mechanics. Action does not transcend Newton in any conceivable way, it only compresses his method. Just as velocity is a compression of distance and time, the Lagrangian is a compression of kinetic and potential energy. Each compression is a mathematical abstraction, because the individual variables are no longer expressed singly. They often do not appear in the equations at all. They are included only a parts of greater variables.

From an engineering standpoint, this is a real advance. As long as the greater variables express the changes of the individual variables in the right way, abstract systems like this can save a lot of time. But from a theoretical standpoint, abstract mathematics can be a great danger. Since the individual variables are no longer in the equations, it becomes much more difficult to see when they are being misused. Abstract mathematics must assume that all its original assumptions are applying with each new application, and with many new applications this may not be so. If time and distance are not behaving in normal ways, then the equations have no way of correcting for that, since they don't have any way to express it. The equations rely on original definitions and assignments, and modern mathematicians and physicists do not usually bother to check to be sure that all these definitions and assignments hold for each new application. They don't do this for two reasons. One, they often don't know what the original definitions and assignments were. The mathematical systems are taught as abstract systems, where foundations are considered to be movable. In the case of the Lagrangian, for instance, we are taught that the variables are general coordinates that we can apply to almost anything. Well, this is true to only a limited degree, and the limits have been ignored. Two, definitions and variable assignments are considered to be metaphysical, and therefore beneath the notice of mathematicians and scientists. Modern scientists cannot be bothered to look at foundational questions, since math is only the equations themselves. If you have mastered the manipulations, you have mastered the math, they think.

To be very clear, it is not action variables I object to. What I object to is their misuse. They are misused when they are applied to systems that do not match the time and distance assignments they were created for. I also object to the implied superiority of action variables. They are very efficient in some uses. But because they are abstract, they are prone to misuse. In this way they are actually inferior. They are inferior because they are less transparent than Newtonian variables. Newtonian variables are not always transparent either, but action variables are *always* less transparent. Action variables are the first cloaking of physics. And in some cases this cloaking is not an accident. Action variables and the math surrounding action is not always used to generate efficient solutions in familiar situations. It is now often used to blanket over holes in theory or math. Like many other mathematical systems, it is now used to mask purposeful fudges.

The next mathematical system that invaded physics is that of Gauss and Riemann, invading through the door of General Relativity. This was really the first major invasion, and the most important. Up until then, physicists had been wary of allowing mathematicians to define their fields, especially with the new abstract systems. The action principle had not yet invaded physics on a full scale, and would not until the arrival of quantum mechanics. Einstein himself was very wary of abstract math, purposely avoiding it until 1912. Put simply, he "did not trust it." But in that year he discovered Gauss, and called on his friend Grossman to help him with the math. A couple of years later Einstein was hired in Berlin, and there he got even better help, from Hilbert and Klein, no less. Einstein had asked the wolf in at the front door.

I don't think it is an accident or coincidence that the first thing the wolf tried to do is take over the house. Hilbert, after schooling Einstein on all the latest techniques, tried to beat Einstein to the punch by publishing the theory of General Relativity two weeks before him. He didn't succeed in this dastardly trick, but amazingly history has not held it against him. Einstein quickly forgave him, and now Hilbert is treated as the greatest mathematician of the 20th century. But to me, this incident perfectly presaged the way the 20th century would go. The mathematics department, invited to consult, would see its opportunity to steal the show, and it has since stolen the show. Someone like Feynman could throw barbs at the math department, but this was only misdirection. The top mathematicians could look back over their shoulder in feigned opposition, only because they had already taken over the physics department. Feynman was not smirking at mathematics, he was smirking at mathematicians who were too narrow to crossover and become famous, like he had. It was as if to say, "We now own physics, the queen of the sciences and the modern kingmaker, and you guys prefer to argue over trivialities like Fermat. That will never win you a Nobel Prize or a trip to the White House."

Einstein's success with the tensor calculus called all the present demons out of the closet, invited them all into the kitchen, and gave them control of the fire. He showed the road to fame, and the first stop on that road was enlisting a new abstract mathematics. That has been the road ever since, and it defines the current low road of string theory, which had planned to awe all opposition with a math so great and so abstract there was no beginning or end to it (its plan is not moving as planned). Quantum Mechanics was the first to learn this lesson, though, and Heisenberg was the greatest student. Heisenberg understood first and best how to use mathematics to impress and cow the audience. He also understood first and best how to use mathematics as a tool of propaganda. A math of proper abstraction and complexity could be used to hide all error, to divert all effort, to deflect all criticism. It could be used like a very heavy, very highly decorated quilt, covering the bedbugs beneath. This new abstract math would come not with a foundation, but with a manifesto. It did not have axioms, it had public relations. It was not sold with an explanation, but with an "interpretation", and this interpretation was to be accepted on authority.

The takeover in the 20th century was very quick once it began. The mathematician Minkowski reworked Special Relativity before the presses had even cooled on Einstein's paper, expressing the field in complex and abstract terms. This reworking was completely unnecessary, but it was accepted just as fast as it was offered. The novelty of it was enough to complete the sale, although the price was steep indeed. The problem with Minkowski's math is the same as the math of action: the danger is all in the loss of transparent variables. Again, I have nothing against complex math as long as it is used with discretion and complete honesty. But Minkowski fails miserably on both counts, as I have shown. The symmetry is a manufactured symmetry, and the loss of the time variable has been disastrous. The subtle errors in Einstein's math were immediately cloaked under an abstract math, and that abstract math was in no way more elegant than the simple algebra of Einstein's original paper. Einstein's paper was dense, but that was Einstein's fault, not the algebra. Minkowski's unstated axioms were not only unstated and unnecessary, they were false. The time dimension does not travel orthogonally to the other three, and this is not a metaphysical subtlety. It is a physical and mathematical fact, easily proved. Even Einstein called Minkowski's math "superfluous erudition." He was only half right. Minkowski's math was certainly superfluous, but it was false pedantry, not erudition. It was sciolism.

Even this unnecessary abstraction and obstruction was not enough to satisfy. Another level was soon added by the tensor calculus, a blanket ten times as heavy as the blanket of Minkowski. Although I have shown that General Relativity can be expressed with Newtonian variables, a Euclidean field, and high-school algebra, the worthies of the time preferred to express it with an undefined curved field and a hatful of unwieldy tensors. In his previous mood, Einstein had said, "You know, once you start calculating (with abstract mathematics) you shit yourself up before you know it." But suddenly, in 1912, he developed a fondness for this mess. Perhaps he saw darkly what Heisenberg would see very clearly: the 20th century would have a love affair with shit. The century proved this in every field, from art to math to science to war to politics to entertainment to sex. The century loved nothing so much as watching someone foul himself in public, as long as that someone could sell the spectacle as a transcendent event.

Once again, a Gaussian field and tensors and all that has followed can be made to work. In some situations it is actually useful. I am not arguing that these fields or manipulations are necessarily false. What I am arguing is that physics doesn't need them. The physical field is not that complex. We have invented maths that are much more complex than we need, and we have gotten lost in their mazes. The problem with the math of General Relativity is that it cloaks the mechanics involved. It is too abstract by several degrees. This means that although Einstein sometimes found a way to get the right answer with all this math, he just as often got the wrong answer. The math is so difficult that almost no one can sort through it and tell when the answer is right and when it is wrong. Even worse is the fact that the opacity of the math makes it impossible to unify it with any other math. The primary events are buried so deep and are so poorly defined that there is no hope of expressing them with the mathematical tools available, or isolating them so that they can be located in other fields. The mathematical manipulations become the primary events, and the mathematical field becomes reality. The math ends up usurping the mechanics.

This opacity causes another problem. Because the primary variables are buried under so many abstract layers, they cannot be studied when problems arise. Later repairs cannot be done at ground level, they have to be done in end-math that adds complexity. In QED this end-math is called renormalization. In GR it is called other things, but in either case it leads to an endless scholasticism and an endless and unsightly tinkering. It ends up providing physics with equations that are post-dictive instead of pre-dictive. Every new experiment requires a new fix, and each new fix is pasted over all the others. You then end up with what we have: a physical math that is burdened with so many fields and operators and manipulations and names that it makes Medieval biblical exegesis look like a cakewalk. And it leads to the absurd situation of having physicists who invoke Occam's razor and the beauty of simplicity offer us a proliferation of fields and manipulations that is truly mind-numbing. When I see a string theorist invoke Occam's razor, I can't help feeling queasy. It is like Fox News invoking honesty in reportage.

Next came Quantum Mechanics. Heisenberg saw Einstein's success with the matrices and voilà, the matrix moved to QM, making it even more famous than GR. But this time we got a confluence of new abstract maths: the turn of the screw. It was feared the matrix would not be enough to wow the world, and so the matrix was joined by the Hamiltonian and Hilbert Space and Hermitian operators and eigenvalues and so on. It was never explained why quanta could not travel in Euclidean spaces under transparent variables, just as it was never explained why gravity required tensors. It was never explained because no one needed an explanation. All were quite satisfied to have new things to do. The new math was the main draw. It gave the theory a required ballast and made everyone look smart. What was there not to like?

Well, there was the fact that everything was based on

probabilities, that the mechanics was contradictory and unfathomable, that many insoluble paradoxes were created, and that the math required an infinite renormalization that was basically "hocus-pocus." But I mean, other than that, what was not to like? If we could just learn to accept that Nature no longer made sense, we would be just fine. After all, the math was big enough to make up for everything. What was Nature next to a math that could fill blackboards? [For a full critique of the math of QED, gauge theory, see the chapter on Weak Interaction.]

As David Politzer, Nobel laureate and inventor of asymptotic freedom put it,

English is just what we use to fill in between the equations.*

Which may explain why the equations have gotten ever longer and the English evermore tenuous and fleeting. Theory must be stated in English—we have no theory— therefore we need no English. Equations will do.

And now that QED is "perfect", we graduate to the even bigger blackboard that is string theory. Since a huge unfathomable math was so successful in QED, string theory naturally developed an even huger and more unfathomable math, one with exponentially more paradoxes and contradictions and ad hoc fixes. If QED requires an infinite renormalization, string theory requires a trans-infinite renormalization. Since QED so successfully ignored mechanics, string theory ignores it even more thoroughly. QED had to state out loud that it was going to ignore mechanics, as a matter of some sort of principle (we are not sure what principle). But string theory goes to the next level of ignorance, which is ignoring that mechanics exists or ever did exist. Like Mephistopheles, the string theorist can call up any entity he likes, just by a simple conjuring. He doesn't need an axiom or a proof or even a definition. All he needs is a need. Science is now defined by desire more than anything else. "I desire a ten-dimensional donut with spikes like a pufferfish, and a gauge theory in the shape of a wombat sitting in the corner smoking a cigar, therefore the universe and this computer model must supply me with one. Oh, and all this exists beneath the Planck limit. Except for the cigar, which disappears in the presence of a scanning microscope."

Yes, modern physics has become a neo-scholasticism. It is the avoidance of real questions in the pursuit of trivial methodology. It is the memorization of an endless list of names and manipulations in lieu of understanding mechanics. It is the setting up in some black data hole and extemporizing on an endless string of evermore ridiculous hypotheses instead of looking at known physical problems closer at hand. It is the knee-jerk invocation of authority and the explicit squelching of dissent. It is the hiding behind tall gates and a million gatekeepers, and euphemizing it as "peer review." It is the institutionalized acceptance of censorship and the creation of dogma. Grand Masters like Feynman say "shut up and calculate!" and everyone finds this amusing. No one finds it a clear instance of fascism and oppression. An internet search on "against Feynman" or "Feynman was wrong" or "disagree with Feynman" turns up nothing. The field is monolithic. It is completely controlled and one-dimensional. All discussion has been purged from the standard model, and all debate has been marginalized. Any non-standard opinion must be from a "crank" and blacklisting is widespread. Publishing is also controlled, both in academia and in the mainstream. Einstein already found science publishing too controlled for his taste in the 30's, refusing to work with *Physical Review*. What would he think now? Can anyone imagine his early papers getting published in the current atmosphere?

If you are an insider at a major university, you can publish anything, the more absurd the better. You can say anything without fear of contradiction or analysis, since science most wants right now to be creative, and it thinks (like modern art) that absurdity is the most creative thing possible. The paradox is the highest distinction, the contradiction the surest sign of elevation. The contemporary physics paper has become like Dubuffet's *La Lunette Farcie*, a purposeful mockery of all convention, a nothing packaged as a something. Soon the physicist may be expected to follow Duchamp, publishing a toilet seat as a TOE.

Contrary to what we are told, contemporary physics is not booming. It is not very near to omniscience, it is not the crown jewel of anything. In fact, it is near death. It has been damaged by any number of things, only a few of which I have mentioned by name here. But the prime murderer has been abstract mathematics. Physics has succumbed to a suffocation. It is the victim of a strangulation. It is in a not-so-shallow grave, and piled on top of it like dirt are a thousand fields and operators and variables and names and spaces and terms and eigenvalues and dimensions and criteria and functions and coordinates and conjugates and bases and bijective maps and automorphism groups and abelian gauge fields and Dirac spinors and Feynman diagrams and so on *ad nauseum*. The only way the grave could be any deeper and darker, in fact, is if we allowed

Deconstruction to dump its transfinite dictionary of onanic terms on top of this one.

The only road out of this grave is to start digging in the upwards direction, clearing away all this schist. The sort of math that physics requires is a math of rigorous definitions and transparent variables, with as little abstraction as possible. We don't need spaces of infinite dimensions, since we don't have infinite physical dimensions. We don't need abstract operators, we need direct representation of motions and entities. Taking the advice of Thoreau, we must "simplify, simplify, simplify." That is our only hope of a Unified Field and a mechanical explanation of the universe.

*http://pr.caltech.edu/periodicals/CaltechNews/articles/v38/asymptotic. html

Chapter 3

The Calculus is Corrupt

$$\frac{}{A \vdash A}\text{ init} \qquad \frac{\Gamma_1 \vdash A, \Sigma_1 \quad \Gamma_2, A \vdash \Sigma_2}{\Gamma_1, \Gamma_2 \vdash \Sigma_1, \Sigma_2}\text{ cut}$$

$$\frac{\Gamma \vdash \Sigma}{\Gamma, !A \vdash \Sigma}\text{ weak}_L \qquad \frac{\Gamma \vdash \Sigma}{\Gamma \vdash ?A, \Sigma}\text{ weak}_R \qquad \frac{\Gamma, !A, !A \vdash \Sigma}{\Gamma, !A \vdash \Sigma}\text{ contr}_L \qquad \frac{\Gamma \vdash ?A, ?A, \Sigma}{\Gamma \vdash ?A, \Sigma}\text{ contr}_R$$

$$\frac{\Gamma, A, B \vdash \Sigma}{\Gamma, A \otimes B \vdash \Sigma}\otimes_L \qquad \frac{\Gamma_1 \vdash A, \Sigma_1 \quad \Gamma_2 \vdash B, \Sigma_2}{\Gamma_1, \Gamma_2 \vdash A \otimes B, \Sigma_1, \Sigma_2}\otimes_R \qquad \frac{\Gamma \vdash \Sigma}{\Gamma, 1 \vdash \Sigma}1_L \qquad \frac{}{\vdash 1}1_R$$

$$\frac{\Gamma, A \vdash \Sigma}{\Gamma, A \& B \vdash \Sigma}\&_{L1} \qquad \frac{\Gamma, B \vdash \Sigma}{\Gamma, A \& B \vdash \Sigma}\&_{L2} \qquad \frac{\Gamma \vdash A, \Sigma \quad \Gamma \vdash B, \Sigma}{\Gamma \vdash A \& B, \Sigma}\&_R \qquad \text{no } \top_L \qquad \frac{}{\Gamma \vdash \top, \Sigma}\top_R$$

$$\frac{\Gamma_1, A \vdash \Sigma_1 \quad \Gamma_2, B \vdash \Sigma_2}{\Gamma_1, \Gamma_2, A \otimes B \vdash \Sigma_1, \Sigma_2}\otimes_L \qquad \frac{\Gamma \vdash A, B, \Sigma}{\Gamma \vdash A \otimes B, \Sigma}\otimes_R \qquad \frac{}{\bot \vdash}\bot_L \qquad \frac{\Gamma \vdash \Sigma}{\Gamma \vdash \bot, \Sigma}\bot_R$$

$$\frac{\Gamma, A \vdash \Sigma \quad \Gamma, B \vdash \Sigma}{\Gamma, A \oplus B \vdash \Sigma}\oplus_L \qquad \frac{\Gamma \vdash A, \Sigma}{\Gamma \vdash A \oplus B, \Sigma}\oplus_{R1} \qquad \frac{\Gamma \vdash B, \Sigma}{\Gamma \vdash A \oplus B, \Sigma}\oplus_{R2} \qquad \frac{}{\Gamma, 0 \vdash \Sigma}0_L \qquad \text{no } 0_R$$

$$\frac{\Gamma, A \vdash \Sigma}{\Gamma, !A \vdash \Sigma}!_L \qquad \frac{!\Gamma \vdash A, ?\Sigma}{!\Gamma \vdash !A, ?\Sigma}!_R \qquad \frac{!\Gamma, A \vdash ?\Sigma}{!\Gamma, ?A \vdash ?\Sigma}?_L \qquad \frac{\Gamma \vdash A, \Sigma}{\Gamma \vdash ?A, \Sigma}?_R$$

The current and historical method for differentiating is a mess. In Lagrange's derivation of the virial, we find him differentiating x^2 with respect to t, to find $2xdx/dt$. He then lets dx/dt equal v, so that dx^2/dt is equal to $2xv$. I found this astonishing, but then was more astonished to find that it is done all the time, to this day. No one calls Lagrange on this cheat because they want to use it themselves. It is the same reason the Democrats never called the Republicans for stealing elections with voting machines. The Democrats wanted their turn with the machines, and got it.

We could continue this cheat of Lagrange to find that $2xv = 2x^2/t$, which would mean that the derivative of x^2 with respect to t is $2x^2/t$.

Does any of this make any sense? Does anyone ever bother to check to see if these equations are physically true? Apparently not.

If the derivative of x^2 with respect to t is 2xv, then what velocity is that? I will be told it is the velocity at (x,t), but 2x is already the velocity at (x,t). Because our power is 2 here, we have a curve equation. The derivative of a curve at a point is a velocity. It is supposed to be the instantaneous velocity at that point on the curve. The derivative is the tangent to the curve, remember, and it is also the velocity at that point. So by this finding of 2xv, we appear to have two simultaneous velocities at a single point, multiplied together. The value 2x is the velocity, and v is also the velocity, so we actually have v^2. The derivative of a power 2 curve equation is a velocity squared?

The derivative of x^2 with respect to t cannot be 2xv, since the derivative is the rate of change. The rate of change of x^2 with respect to t cannot be 2xv, unless v=1. Historically, Newton used the notation dx/dt (or the equivalent) here simply as a reminder of the relationship. It is like a listing of physical dimensions in an equation, not a continuation of variables. For example, if we find the derivative of x^2 with respect to x, we find 2x, which could be written 2x(dx/dx). That would be read, "The rate of change of x^2 is 2x(x with respect to x)." But since it is clear that dx/dx=1, we drop the notation.

When we find the derivative of x^2 with respect to t, we are doing the same thing. We can write that as 2x(dx/dt), but only if we remember that dx/dt is just notation. In that equation, dx/dt is reducible to 1, so it cannot be written later as a velocity *variable*.

In fact, in this notation, dx/dt IS the dimensional notation. It does not stand for a velocity variable, it stands for meters/second, or something like that. If we find the derivative of x^2 at x=4, the answer is 8, and the dx/dt only stands for the dimensions. We need dimensions, since 8 is not a physical answer. Since x^2 is an acceleration, the rate of change of that curve at a given x must be a velocity. The derivative of an acceleration is a velocity. So dx/dt is giving us the velocity dimensions of length over time. The ratio dx/dt is not a velocity variable, it is dimensional constant.

The ratio dx/dt is reducible to 1 simply because, physically, distance and time cannot be changing at different rates, as infinitesimals or at the limit. You will say that if our velocity is 2, our distance is changing at a rate of 2 while our time is changing at a rate of 1, but in the calculus, first order changes like that are ignored. The derivative of 2 is 0 and the derivative of 1 is 0. That is why the ratio in the notation is written dx/dt. It is not written x/t. The letter d here does not mean delta, it means fluxion or infinitesimal change. At the limit, all first order changes are equal, so dx/dt is vanishing in the same sense that 1 is vanishing in equations. It is vanishing in the sense that you can ignore it. You can cross it out of the equation. If you were Newton, you would think of it as a dot over a dot, and let it vanish for that reason. If you are with me, you think of this as 1/1, which is "vanishing" for a different but mostly equivalent reason.

That notation dx/dt here is not a representation of how x and t are changing relative to each other in any *specific* problem, it is a representation of how x and t are changing relative to each other in the physical field or space, absolutely. It is analogous to dx/dy. If we study space itself, does x change at a different rate than y? Do we measure the width of space differently than we measure the depth of space? Does space itself get wider faster than it gets longer? No. In physics, dx/dy always equals 1, just as dx/dx always equals 1. For the exact same reason, dx/dt always equals 1 *in this sort of notation*. Since time is defined relative to distance, they cannot be changing at different rates. If x is changing at a rate of 1, t must be changing at a rate of 1, and this has nothing to do with the velocity of any real object in any real measurement.

To show what I meant when I said that this notation is equivalent to dimensional notation, we can look at some common equation in physics. Say,

$G = 6.67 \times 10^{-11} \ m^3/kgs^2$

You can't rewrite that as

$G = 6.67 \times 10^{-11} \ v^2m/kg$

And then start inserting different values for v into that equation. That second equation is true only in the case that v=1. v is not a variable there, it is just a dimensional analysis.

It is the same with the notation dx/dt. The ratio dx/dt can be written as a velocity only if you remember that v=1. In that case, v is not a variable, it is just a constant dimension, reminding us of a prior relationship. In Newton's and Leibniz's notation, they used dx/dt (or its equivalent) only in this way.

Therefore, Lagrange's math is just a cheat. Any time you see dx/dt written as a velocity, in a situation like this, a big red flag should pop up.

You will say, "C'mon, we all know that dx/dt is a velocity. What are you talking about? Velocity is *defined* as dx/dt, for heaven's sake!" Yes, in many situations, it is. I am not denying that dx/dt is a velocity, as long as it is used correctly. What I am denying is that dx/dt *in this particular case* is equivalent to the definitional notation of velocity. I am pointing out something fundamental and of great importance, and you better open your eyes to it. The notation of calculus has always been convoluted and sloppy, and this sloppiness had already reached epidemic proportions by the time of Lagrange. If you differentiate x^2, finding $2xdx/dt$, the dx/dt in that notation is not a velocity. The notation is telling us that we are finding the rate of change of x^2 with respect to the given rate of change of time. Since the given rate of change of time is always 1, dx/dt must also be one.

Let me put it another way. I try to explain this in as many ways as possible. In physics, does time ever change at any other rate than 1? Can time itself be accelerated or dilated? No. Not even in Relativity. In Relativity, *measurements* of time can be dilated or compressed, but time itself is invariable. We never find time to any power. We never find time going at any rate above 1 or below 1. Time is always ticking at a rate of 1, by definition. The rate of change of time is and must be 1. Therefore, if we are given the ratio dx/dt, and we know that dx is 1, then dx/dt must be 1. Well, if we differentiate some power of x with respect to t, then dx can hardly be anything but 1, can it? Given x^2, we assume dx=1. If we differentiate x^2 with respect to x, we assume dx=1, don't we? If we didn't, then we couldn't find that $dx^2/dx=2x$. The denominator dx has to equal 1 or the equality is ruined. In the same way, when we differentiate with respect to t, dx is still changing at a rate of 1. Therefore, dx/dt=1. It can be dropped.

If we differentiate x^2 with respect to t, the answer is 2x, not 2xdx/dt. This sloppy and confusing notation should be abandoned. The ratio dx/dt should mean one thing in calculus and one thing only. It should not be used differently in different places, since that only encourages this sort of cheating.

I have also seen x^2 differentiated with respect to t to get 2xx'. That is unnecessary for the same reason. It just gives cheaters another variable to play with, to fudge later if they want. The term x' just means the derivative of x, and of course the derivative of x is just 1. So there is no reason to write it. The only reason to write it is so that you can use the Lagrange fudge later, claiming that x' = v, and push your equations that way. I have seen "real" mathematicians defending this manipulation, claiming that whenever you differentiate a length, you get a velocity. But that the upside-down calculus I have talked about elsewhere. Many mathematicians and physicists actually believe they can differentiate any length they like into a velocity. But they can't, as I have shown with Lagrange and his virial proof. They have to be given an acceleration, and every length is not part of an acceleration. Because they have misunderstood calculus from day one in high-school, they think that when they differentiate some power like x^2, they are differentiating a length. But they aren't. When you differentiate x^2, you are differentiating a curve, which is an acceleration. The term x^2 is *already* an acceleration, even without a ratio, a denominator, or a "with respect to." The term x^2 stands for this accelerating series of numbers:

x^2 : 1, 4, 9, 16, 25, 36, 49, 81

Therefore, you differentiate an acceleration into a velocity. You do not differentiate a length into a velocity. You differentiate down, not up.

But as it is, mathematicians think they can differentiate both up and down. They can differentiate x^2 into a velocity, because $2x$ is a velocity at the point x; and they can also differentiate x into a velocity. Since x' = v. But their second manipulation here is illegal, since they have just differentiated *up*, with no given acceleration or curve. If you are given x, you are not given a curve or any variable change. You are given this series of numbers:

x : 1, 2, 3, 4, 5, 6, 7, 8

The derivative of that series of numbers is 1, because the rate of change of that line is 1. There is a difference of 1 between each and every term. If you assign a velocity to that series, it is also 1. You cannot differentiate x into a velocity *variable*. You can only differentiate x into the velocity 1.

Therefore, the calculus has been corrupt and almost infinitely fudgable, at least since the time of Lagrange.

45

Chapter 4

The

Virial Theorem *is* False

The virial theorem comes from the math of Lagrange, as extended by Clausius. The historical and current derivation starts with the moment of inertia, so the virial must apply to a collection of particles arrayed about a center. That is, it applies to circles or spheres.

$$I = \Sigma mr^2$$

That is a summation of all the individual moments of inertia of N number of "point" particles in our collection. Notice that m applies to the mass of each point particle. Our first question here is, "How can a point have mass?" It can't, but we will move on, since we have much more important and obvious fudges to show. The virial G is then defined as the sum of the motions of these masses:

$$G = \Sigma mvr$$

That expression looks like angular momentum, L = mvr, but it isn't. Beware. The v has not been defined as a tangential velocity. The v there is found by differentiating r, and you cannot differentiate r to find a tangential velocity. Rather, Lagrange, the inventor of this derivation, then relates the two equations:

$$G = dI/2dt = \Sigma mrdr/dt$$

As you see, he has differentiated one of the r's and left the other one standing. Not satisfied with making a distance into a velocity, he then "differentiates" again:

$$dG/dt = \Sigma pdr/dt + \Sigma rdp/dt = \Sigma m(dr/dt)(dr/dt) + \Sigma Fr = 2K + \Sigma Fr$$

Where K is the summed kinetic energy of the system. You can see that he has "differentiated" twice to turn a distance r into a velocity squared. That allows him to to end up with a force F at r and a kinetic energy.

But let's unwind that mess. It is a horrible fudge from beginning to end. We start by noticing that r is just a simple distance in the first equation. It is the distance from the center of some "point" particle among N point particles in our rotating system. Lagrange wants to turn that distance into a velocity, so that he can create a momentum. He needs mv, you see. So he takes the "time derivative" of r. Not of r squared, as you see, but of just *one* of the r's he has in the first equation. By this magic, the r becomes a v.

I have said that calculus is misunderstood and misused, and this is perfect example. I entered this problem thinking I was only going to have to make a correction based on my correction to the angular momentum equations, but my work on the calculus is paying off, because I can now show that this is fake calculus. Lagrange is pretending to do calculus, but he is just finessing equations. Amazing that no one has had a problem with that for 230 years.

When you actually do real calculus and you find a real derivative, you are differentiating a curve or an acceleration. You are finding the rate of change of that curve or that acceleration. But you can't find a rate of change of a distance r, since there is no change. In other words, you can't differentiate r with respect to t, since r is not changing with respect to t. Just think about it physically for half a moment. The distance r is defined as the distance of point k from the center at time t. Now let some time pass. Has the distance r changed? The current derivation has nothing to say about it, but we can't assume that r is moving, much less accelerating, since that would mean all our particles in our system are swapping places at a fantastic rate. Maybe they are, maybe they aren't, but this system has not been defined as a gas, remember. The virial theorem is now used on stars, dark matter, white dwarfs, and so on, and in stars there is no reason to believe that matter in the star is changing its radius over time. Is the core of our Sun accelerating out into the corona, or vice versa? And if it were, wouldn't all this motion have to offset? All the motions of all particles k would have to sum to zero, or the star would be exploding, would it not? In other words, the distance from center r cannot be changing, as a sum. If our system were a gas, the same would apply: the total change in r over time could not increase, or our gas would be expanding very quickly.

This is all just to say that r cannot be taken to be changing with time. Lagrange has just buried the mechanics here, and the math is unsupported. Unless he can show that the distance of every particle is increasing to some power with time, he cannot differentiate r.

Besides, in the moment of inertia equation, any velocity is a tangential velocity. Just look at the definition of the virial. The virial looks just like a summed angular momentum, in form. The angular momentum L is equal to mvr, and so is the virial. Well, the velocity in the L equation is a tangential velocity. In circular motion, the value of the velocity does not change over time, since the tangential velocity is determined by the distance r. If r is a constant, then v is a constant. I have just shown that we cannot be letting r change with time, either as a velocity or as an acceleration; therefore, v cannot be changing either.

You can't turn any distance you like into a velocity just by finding some nebulous "time differential". You can differentiate a distance into a velocity only if you have a variable velocity: an acceleration. If your velocity is either constant or zero, you cannot differentiate a distance, since there is no variance. There is no change with time to study or represent with math. In the case of the virial, we have the appearance of a tangential velocity which is constant, but even that is not what the equation is representing, as I warned above. The first equation is written with respect to r, *which is not changing at all*. Therefore, the "velocity" of the distance r is zero. We have no velocity here at all. The math above is just magic.

But even that isn't the whole of the fudge. In order to differentiate the r in the first equation, this derivation has written r^2 as r dot r, and then differentiated only one of the r's. But r^2 is not a dot product, since I and r are both scalar.

The formal derivation at Wikipdeia defines *I* as scalar, so it is no bold claim of mine. If we go back in time to find where r^2 comes from, we find that it comes from this equation: $v = r\omega$. Again, that r is just a distance, not a vector. When we square it, we are not finding a dot product, we are just squaring it because the v is squared in the deriving equation:

$$K = \frac{1}{2}\,mv^2 = \frac{1}{2}\,m(r\omega)^2 = \frac{1}{2}\,(mr^2)\omega^2 = \frac{1}{2}\,I\omega^2$$

See, r is not a dot product, it is just a squared variable. Another way to see this is to imagine that r *were* changing with time. If r were changing with time, we would find r squared by a straight squaring, as if it were in a gravitational equation or something. We wouldn't have to concern ourselves with vectors or dot products. By definition, this is due to the fact that the radius itself never has an angle to the center. Two *positions* of a particle may have an angle to the center, but the *distance* does not, since the distance always has its origin *at the origin*.

For this reason, you cannot possibly differentiate r with respect to t, finding rdr/dt. If we differentiate *I*, then it will be r^2 that is changing with respect to time, not r. In which case, the derivative of r^2 is 2r. But Lagrange didn't want 2r, he wanted a velocity, so he simply fudged his way into it.

And there is even more fudge. Every line is a new cheat. Look here:

$dG/dt = \Sigma pdr/dt + \Sigma rdp/dt = \Sigma m(dr/dt)(dr/dt) + \Sigma Fr$

In the last term, we find that F=dp/dt. Funny that Lagrange

never has to concern himself with angular momenta here, even though we are in a circle or sphere. His virial G is written as a summed angular momentum, but he and the historical derivation don't want you to notice that. I have pointed it out, but textbooks never do. To have a moment of inertia, you have to have some rotation, so we cannot imagine that our collection of points here is just static, arrayed around a center. If that is so, then why is the virial expressed with p and v, instead of L and ω? F=dp/dt is a linear equation, not an angular equation. In an angular solution, we would have T = dL/dt. We would have a torque.

The reason this derivation ignores all this is that it wants you to think the velocity in the equation $G = \Sigma mvr$ is defined in some way, when in fact it isn't. The virial equation doesn't write the angular velocity as ω, because if it had you would have recognized for yourself what I just recognized. You would have seen that you can't differentiate a radius r into a ω. No, dr/dt ≠ ω . The virial equation is given us right after the equation $I = \Sigma mr^2$, so we think the v is like the velocity in L=mvr. But as I have shown, the v is actually obtained by differentiating r, and r implies no motion of any kind, neither a tangential velocity nor an angular velocity, nor any other change of r.

Also, the v in the equation L=mvr is not found by differentiating r, as you will realize if you think about it. In that equation, it is clear that change-in-r cannot be causing v, since r is not changing. Historically, what is thought to be causing v is an "innate motion." Or at least that is what Newton told us. He assigned the tangential motion to an independent linear motion, and that motion became curved by the influence of a centripetal acceleration acting upon it.

This must mean that differentiating r to find a velocity is a double and triple fudge, worthy of the all-time hall of fame. Lagrange has pushed this derivation in every single line.

I came in expecting to correct the equation L=mvr, but that equation never raised its head. The virial derivation had to drive right around L, because the virial is just a variation of L, and no one wanted you to realize that. If the virial can be written as a sum of kinetic energy and torque, why can't the angular momentum be written that way? If we sum L and then differentiate, will we find the same thing? No, because the v in L=mvr is defined as the tangential velocity, and the v in the virial is undefined. That is also why the force at r can't be defined as the torque: in the torque equation, the velocity is defined as the tangential velocity. In the virial, the velocity is undefined. They don't want to connect Fr to the torque, because they want to connect it to the potential energy.

All this is swept by very fast in the derivation, since historically they want to bury that velocity as quickly as possible. The virial is now commonly written without any mention of a velocity, as in Lagrange's identity:

$$dG/dt = d^2 I / 2dt^2 = 2K + V$$

where V is the potential energy of the collection. Which of course leads us to the Lagrangian. But this means that the virial and the Lagrangian are meant to be applied to gravitational systems. That is why they are interested in potential energy rather than torque. That is why spin is ignored: they want to beable to apply the equation to gravity fields, spinning or not.

That, of course, confirms my initial analysis. If these energies are the result of a gravitational field, then the distance r of particle *k* from the center cannot be accelerating over time. The particle *k* may be *feeling* an acceleration, but it cannot be accelerating, short of some revolutionary new postulate. For instance, I could be defined as some particle *k* in the field of the Earth, since, relative to the Earth, I am pointlike enough. But my radius r is constant. It cannot be differentiated with respect to time, can it? If we differentiate my position r with respect to time, we will find pretty much zero. We could say the same of most objects in the Earth or in its field. Only objects in freefall could fit into the virial equation. Since a star is not a collection of objects in freefall, I do not see how we can apply the virial equations to a star.

The virial equation and Lagrange's identity have also suffered from another big disclarity. Physicists have always wondered where that 2 comes from. Why don't the kinetic energy and the potential energy offset exactly, as they should? Why is the potential energy twice the kinetic energy? If we look at the definitions of potential and kinetic energy, the two should offset. One should be the opposite of the other, and all we should need is a minus sign to create the equality. By the work-energy theorem, kinetic energy is Fd. By the definition of gravitational potential energy, we find V=mgh. Since F=mg, this becomes Fh.   The variables h and d are equivalent, since in one it is the distance the object travels, and the other it is the distance the object *would* travel, if released. So we should have an equality. If the potential energy is twice the kinetic energy, then why doesn't the object in the field have half its energy left after it falls to center? I will now show you the cause of this gigantic error.

An even simpler derivation of Lagrange's identity goes like this:

$$-V = GMm/R$$
$$F = GMm/R^2$$
$$C = ma = mv^2/R$$
where C is the centrifugal force
$$GMm/R^2 = mv^2/R$$
$$K = mv^2/2 = GMm/2R$$
$$2K = -V$$

The problem there is in the third equation. I have shown that $a \neq v^2/R$. Instead, $a = v^2/2R$. Yes, unbelievably, Newton made a simple error in his versine proof, and that error has never been corrected until I found it. It has been "confirmed" by all physicists and mathematicians since Newton, and re-expressed with modern calculus. However, the calculus is forced to fudge the equation in order to make it match Newton's equation.

If we make the correction here, we find that $K = -V$, as it should. There should be no 2 in the virial, and no 2 in Lagrange's identity. In this way, we can see that the proof has been pushed in yet another way. The longer derivation at the beginning can't be correct, since we now know it was pushed in order to find that 2.

Now let us make the other corrections to the virial. The virial equation can't be derived from the moment of inertia, because we want to apply this to a gravity field that is not spinning, if we like. With no spin, you have no moment of inertia. A critic will say that the standard model is not bothered by the requirement of spin, since all stellar and quantum objects seem to be spinning. But that is not the question, is it? The point is that in a gravitational field, neither the kinetic energy nor the potential energy needs to be caused by spin. Neither Newton's nor Einstein's equations require spin to express gravity. Therefore, we should be able to develop a virial equation with no spin. We could add spin later, if we like. Besides, the current derivation says nothing about spin, and the spin is never included in the equations. If we had spin, the current virial equation would have to include a variable for it. It does not. If we have no variable for the rate of spin, then we cannot possibly derive the equation from the moment of inertia. Remember, the v in the virial is not the rotational velocity or the tangential velocity. No, it must apply to some radial change in r. Therefore, the virial contains no spin. If it contains no spin, it cannot have a moment of inertia and cannot be derived from a moment of inertia.

We really have to start over from scratch. The virial theorem is said to find (Wiki), "a general equation relating the average over time of the total kinetic energy of a stable system bound by potential forces with that of the total potential energy." We see in that definition that we are dealing with a gravity field, since what other object arrayed around a center is "bound by potential forces"? To find a summed kinetic energy in such a field, we have to assign a velocity. But if the field is not spinning, what is this velocity? The virial theorem cannot be measuring molecular or atomic motion, since those motions are not mainly caused by the gravity field and would not be balanced against gravitational potential energy. We are told that the virial theorem can be *extended* to include E/M fields, which must mean that the original equations did *not* include those fields.

Therefore, the forces between particles are gravitational in the first instance. What motion in the field does gravity cause? Unless we are talking about orbiting or freefall, it causes none. Since a gravity field requires no tangential velocity, orbiting is not required here. And since we are not told that all objects are in freefall, we must assume otherwise. The virial should be and is applied to particles that make up a star or planet, and the particles that make up a star or planet are neither in orbit nor in freefall.

This means that without the equivalence principle, no velocity can be assigned. So let us propose that gravity does cause real acceleration in all particles in the field. We can believe that is due to real expansion, or we can believe that is due to a vector flip and some mathematical imagination. It doesn't really matter. This sort of math is used all the time now by the mainstream, as I have shown in my Pound-Rebka paper. You cannot calculate gravitational blueshifts without assigning a motion to the surface of the Earth. For proof, I send you once again to *Feynman Lectures on Gravitation*, lecture 7.2.

Either way, we assign the acceleration to r. We start with the common equation $v = \sqrt{(2a_r r)}$. But instead of imagining this equation is caused by the particle falling toward the center, we imagine it is created by the particle being accelerated out from the center. At that acceleration, the particle would acquire a final velocity of v at distance r. The kinetic energy of the particle at r will then be $mv^2/2$, and the summed energy of all particles will be

$K = \Sigma m a_r r$

Since F=ma, that becomes

$$K = \Sigma Fr$$

Now, the virial derivation finds that ΣFr is the potential energy of the system by actually deriving it and assigning it, but that is just one more fudge. Remember that they found

$$dG/dt = \Sigma pdr/dt + \Sigma rdp/dt$$

Then they assigned the first term to kinetic energy and the second term to potential energy. How is that even remotely logical? The first term contains a momentum and a change in r. The second term contains a distance and a change in momentum. Just because you have a force doesn't mean you have a potential energy. In my own derivation, I got a kinetic energy from a force, as you see. And the standard model derives kinetic energies from forces all the time, as in the work-energy theorem, Fd = K. Force times distance equals kinetic energy, not potential energy. You will say that is force *through* a distance, while we are looking at force *at* a distance, but after my analysis, the difference is not so clear. If the particles must be given a motion in the field, then it would appear we actually have a force *through* a distance here as well. Which means that the historical assignment is just arbitrary and willful. They assigned Fr to potential energy for no other reason than that they wanted to. But if this entire historical derivation was derived from some *motion* at r, and since motion is kinetic by definition, then dG/dt must be kinetic *in toto*. Again, they derived all this from the moment of inertia, and the moment of inertia, though a scalar, cannot be static or potential. Their entire derivation depends on motion, hence the time differentials:

you cannot have time differentials without motion. Therefore, they cannot assign any part of dG/dt to a potential.

The current derivation tells us that, "The total force F on particle k is the sum of all the forces from the other particles j in the system," and then uses that definition of force to prove that ΣFr is the potential energy. But it only proves that IF the F in ΣFr is "the net force on that particle," then V is the sum of the potentials. Unfortunately, it is never proved that dp/dt is the net force on particle k. The derivation simply states it. In fact, it is *impossible* that dp/dt in that equation is the net force on particle k, since the equation contains another term, and the terms are added. Once again,

$G = \Sigma$pr
dG/dt = Σpdr/dt + Σrdp/dt

As you see, we already have a sum of momenta in the first equation. Then we differentiate, which gives us the two terms. So we cannot separate the second term from the first. Forces are contained and implied by both terms, and you cannot ignore the force in the first term. In other words, it required some force to create that kinetic energy pdr/dt. K = Fd. If we have a kinetic energy in the first term, we have some force that created it. The two terms are added, so you cannot pretend that all the force is in the second term!

No, we have to find the potential energy in a different way, just as with the kinetic energy. This historical proof is comprised in every part, in every possible manner.

The potential energy is easy to derive without the virial,

since we just sum the individual potential energies. The potential energy of a single particle is mgh, with h being equal to r here. So the sum is

$$V = \Sigma mgr = \Sigma Fr$$

The potential and kinetic energies are equal, as they must be, and we don't require big fudged equations to show it. We especially don't need to derive any of this from a moment of inertia, since a moment of inertia requires spin and a gravity field requires no spin.

What does all this mean for physics? Well, it is complete knock-out, nothing less. Because I have shown that the virial is not only derived by a big fudge but is actually false as well, the very sky falls. As just one example, Fritz Zwicky has used the virial to deduce the existence of dark matter. But I have shown the equation is 100% wrong. Kinetic energy is actually twice what we think it is, or 100% more than the equation tells us, so all cosmological estimates are off by huge margins.

The same can be said for the Chandrasekhar limit for the stability of white dwarfs. The virial radius is also compromised by a large margin. Everyone who has ever used the virial, including Rayleigh, Poincare, Ledoux, Parker, Chandrasekhar, Fermi, Pollard, Schmidt, and thousands of other top names, are not only shown to be wrong, but to be very poor mathematicians. How is it that no one has thought to check the postulates and manipulations of Lagrange for two and half centuries?

The fall of the virial also topples the Lagrangian and the

Hamiltonian, which rocks QM and QED to their cores. The Lagrangian is defined as $L = K - V$. If Lagrange and all subsequent physicists have miscalculated the number relationship of K and V, then L cannot be correct. K is half what it should be relative to V, which makes this equation wrong by a huge margin. And if K and V are equal to one another, then the Lagrangian is simply $L = 2K = -2V$. This falsifies the Lagrangian for Newtonian gravity, which falsifies Gauss' law for gravity. It also falsifies both the Hamiltonian and the action. Basically, we have to rewrite all of physics for the last two centuries.

Some will think that QED can remain clean of this mess, because in QED our field is not gravity. Unfortunately, the virial is thought to be applicable to E/M, and the derivation is much the same. In other words, that false 2 is carried over into the E/M equations via Lagrange's identity. Remember that the virial theorem has been generalized into many forms, including a widely used tensor form. But even in its simplest form, it is thought to be generalized. The virial theorem, as derived above, is not limited to gravitational fields. It is generalized to apply to "any stable system of potentials," and the potentials of E/M are included in that generalization.

Chapter 5

UNLOCKING
the
LAGRANGIAN

The Lagrangian is perhaps the most important bit of math in current physics, since it props up both celestial mechanics and quantum mechanics. In quantum mechanics, the Lagrangian has been extended into the Hamiltonian. The

63

Hamiltonian does nothing to correct the Lagrangian, taking it as true and given. Therefore, any new information about the Lagrangian must have far-reaching consequences for physics, at all levels. Here, I will not only be able to unlock the Lagrangian, showing what mechanics it really contains, I will also be able to show that it is actually false in many situations.

At its simplest, the Lagrangian is just the kinetic energy of a system T minus its potential energy V.

$$L = T - V$$

At Wikipedia, we are told this:

The Lagrange formulation of mechanics is important not just for its broad applications, but also for its role in advancing deep understanding of physics. Although Lagrange only sought to describe classical mechanics, the action principle that is used to derive the Lagrange equation is now recognized to be applicable to quantum mechanics.

What I will show is that the Lagrangian, rather than advancing a deep understanding of physics, actually blocked an understanding of the real fields involved. Because Lagrange (and Hamilton) mis-assigned the fields or operators, and because this formulation has been so successful and authoritative, many generations of physicists have been prevented or diverted from pulling this equation apart. What do I mean by that? Well, if we take Lagrange at his word, we would seem to have only one field here. In celestial mechanics, the gravitational field causes both the kinetic energy and the potential energy. In quantum mechanics, charge causes both the kinetic energy and the potential. But let's start with celestial mechanics, since that is where the Lagrangian initially came from. The motions of celestial bodies are gravitational, we are taught, and the potential energy is gravitational potential. That being so, the Lagrangian must have originally been a single field differential. In other words, we are subtracting a field from itself. Our first question should be, is that even possible? Can you subtract gravity from itself, to get a meaningful energy? Or, to be a bit more precise, can you subtract gravitational potential from gravitational kinetic energy? That would be like subtracting the future from the present, would it not? Potential energy is just energy a body would have, if we let it move; and kinetic energy is energy that same body has after we let it move. So how can we subtract the first from the second?

Another problem is that for Newton, the two energies would have to sum to zero, by definition. This is clear for a single body, and a system is just a sum of all the single bodies in it. Therefore, both the single bodies and the system of bodies must sum to zero, at any one time, and at all times. In fact, Newton actually used this truism to solve other problems. He let potential energy equal kinetic energy, to solve various problems. But here, we are told that potential energy and kinetic energy don't sum to zero, and aren't equal, otherwise the Lagrangian would always be either zero or $2T$. A Lagrangian that was always zero would be useless, wouldn't it, as would a Lagrangian that was just $2T$.

We can see another problem in this quote from Wiki:

For example, consider a small frictionless bead traveling in a groove. If one is tracking the bead as a particle, calculation of the motion of the bead using Newtonian mechanics would require solving for the time-varying constraint force required to keep the bead in the groove. For the same problem using Lagrangian mechanics, one looks at the path of the groove and chooses a set of *independent* generalized coordinates that completely characterize the possible motion of the bead. This choice eliminates the need for the constraint force to enter into the resultant system of equations.

The problem there is that one solves by ignoring forces, looking only at the path. Why is that a problem? Because if you are studying the path and not the forces, you will come to know a lot about the path and nothing about the forces, which is what we see in current physics. The Lagrangian calculates forces by ignoring forces. It goes right around them. If that were just a matter of efficiency, it might be tenable, but we have seen that historically, the Lagrangian and action were chosen to avoid the questions of forces, which physicists were not able to answer. They weren't able to answer them in the 17th century and they aren't able to answer them now. So they misdirect us into equations that "summarize the dynamics of a system" by ignoring the dynamics of a system. Dynamics means forces.

Yes, we are told at Wiki that the Lagrangian is "a function that summarizes the dynamics of a system." So here is yet another problem. We are then told that T is the kinetic energy of the system. Well, shouldn't the kinetic energy already be a function that summarizes the dynamics of the system? Dynamics means motions caused by forces, so the motion of the particles should be an immediate measure of *all* the forces on them. In other words, the gravity field should already be causing motion, so there is no reason to add or subtract it from the kinetic energy. Either the gravity field is causing motion, or it isn't. If it is, then it should be included in the kinetic energy. If it isn't, why isn't it?

But physicists have never bothered themselves with these logical questions. Why haven't they? Because they found early on that the Lagrangian worked fairly well in many situations. Like Newton's gravitational equation, it was an equation that they were able to fit to experiments. This is very important to physicists, for obvious reasons. But the fact that the Lagrangian worked meant that the kinetic energy and potential energy did not sum to zero, which meant that the bodies were not in one field only. To express energy as a differential, you must have two energies, which means you must have two fields. One field can't give you two energies at the same time. You cannot get a field differential from one field. As soon as the Lagrangian was found to be non-zero, physicists should have known that celestial mechanics was not gravity only. It *had* to be two fields in vector opposition.

By the same token, as soon as the Lagrangian was discovered to work in quantum mechanics, the physicists should have known that QM and QED were not E/M only. The non-zero Lagrangian is telling us very clearly that we have two fields. Just as gravitational potential cannot resist gravitational kinetic energy, charge potential cannot resist charge. Charge potential is not charge resistance, it is future charge. You cannot subtract the future from the present in an equation! This proves once again that gravity is present in a big way at the quantum level. I have proved that in other papers, but we should have known it just from the form of the Lagrangian.

The next question physicists should have asked is this: "Given that the Lagrangian is non-zero, and that it works pretty well, what can we infer from that?" Just from the form of the Lagrangian, we can infer that we have two fields, in vector opposition, one field larger or smaller than the other, or changing at a different rate. We can infer these things, because logically they must be true.

What this means is that the Lagrangian was an accidental and incomplete expression of the unified field. *The Lagrangian is a unified field equation.* I have already shown that Newton's gravitational equation was a unified field equation, and that Coulomb's equation was a unified field equation, and it turns out the Lagrangian is just one more unified field equation. Yes, both of the operators is are mis-assigned or mis-defined. The only reason the Lagrangian works is that the operation works, but it turns out the operation works only because of a compensation of errors. The equation has to be pushed to work.

The Lagrangian has sometimes been interpreted as the total energy of a field, so that it really *is* like adding the future to the present. The kinetic energy is energy the particle already has, the potential energy is energy it soon will have, therefore the Lagrangian is an expression of the total field present at a given location. If we want to know where a system is heading, we add its current state to its potential, right? Sounds feasible, but that isn't what is happening. The Lagrangian isn't a sum of present and future, it is a sum of energy due to charge and energy due to gravity. As with Newton's gravity equation, the Lagrangian already includes both fields. We can tell this just because the Lagrangian includes V, and V is a restatement of Newton's gravity equation. Since V is already unified, L must be as well. L is not a unification of present and future, L is a unification of charge and gravity.

So what is T, by this analysis? T is a unified field correction to V, since V doesn't contain enough information to solve. In my unified field papers, I have shown that although Newton's equation is fundamentally or roughly correct, it doesn't contain enough degrees of freedom to solve most real problems. It contains G, which tells us the scale between the two fields, but it doesn't tell us how the two fields vary by size. Newton's equation doesn't include the density of the charge field, which is relatively small at the macro-scale, but more important at the quantum level. In other words, because the photon has real size, it begins to take up more space at the quantum level. This makes the E/M field relatively stronger at smaller scales. It is a larger part of the whole at that level, and a smaller part of the whole at our scale. But Newton's equations have no way of including this information. The Lagrangian is an improvement, because T goes some way in solving this problem. I don't know that Lagrange or Hamilton meant to correct Newton in that way —I suspect they didn't—but the Lagrangian, on purpose or by accident, expresses this degree of freedom. This is because the "kinetic energy" term T includes the mass again. Not only that, but it tells us how that mass is velocitized by the fields present. *We get the mass right next to its own velocity.* Indirectly, this must tell us how that mass is responding to the photon density, which tells us how the gravity field and charge field are fitting together in this particular problem. So the variable T corrects the variable V, giving us a total field energy L that is an improvement on any energy Newton could find or predict.

But why is the Lagrangian sometimes wrong, as I say? Because when you have an equation that is in a confusing and unknown form, it is quite easy to plug the wrong information into it. In its current form, the Lagrangian is potentially useful, since if you do everything right, it will work. But since most or all physicists don't know how or why it is working, they end up plugging the wrong numbers into it.

We see this in the two-body central force problem, where the Lagrangian is used to make a hash of the problem. This is apparent at Wiki from the first sentence, which begins,

The basic problem is that of two bodies in orbit about each other attracted by a central force.

In the two-body problem, two bodies are *not* in orbit about each other. One body is orbiting the other body. Is the Earth orbiting the Moon? No. The Earth may or may not be orbiting a barycenter, but in no case is the Earth orbiting the Moon. Also, in the two-body problem, are the bodies attracted by a central force? No. Each body is attracted by the other body. There is no central force. The barycenter, even were it true, would be mathematical only. No force comes from there. We have seen this sort of language in many other places, and I always find it a bit shocking. How can physicists use such sloppy language? Actually, it goes beyond sloppy, since it is demonstrably false. This language is being used as more misdirection. It is used as a purposeful confusion, so that the reader cannot make sense of anything on the page. But the problems are not just problems of semantics or propaganda, they are mathematical, for we are then given the equation

$$L = T - V = \tfrac{1}{2}\,M\dot{R}^2 + [\tfrac{1}{2}\,u\dot{r}^2 - V(r)]$$

70

Where M is the combined mass, $\dot{R}$ is the velocity of the barycenter, u is the reduced mass, and $\dot{r}$ is the change in distance between the two bodies (the velocity of the separation). That is a hash for so many reasons. One, if we put M and $\dot{R}$ next to eachother in an energy equation, they have to apply to the same thing. One can't apply to combined mass and one to the barycenter. No, M must be the center of mass, not the combined mass. This means we MUST put the combined mass at the center of mass. But if we do that, then we can't have any separation, and if we don't have any separation, we don't have $\dot{r}$. The same thing applies to u and $\dot{r}$. To put them together in an energy equation, they have to apply to the same thing. One can't apply to one thing and the other to another. Therefore, $\dot{r}$ should be the velocity of the reduced mass, not the change in separation. But since the reduced mass is a quotient over a sum, it can't have a velocity. And, since I have shown that reduced mass is a figment from the beginning, it can't be put into any equation. It is false, so it necessarily falsifies any equation it is in.

But it gets even worse. Study that equation some more, and you see that it has not one but two kinetic energies in it. I thought the Lagrangian was already a summation, applying to a system, so how can you justify putting two kinetic energies in there? Shouldn't a system have only one total kinetic energy? It looks to me like (from the brackets) that we are being told that

$$V = -[\tfrac{1}{2}\, u\dot{r}^2 - V(r)]$$

Does that make any sense? Not really, because we are then told that

$L = L_{\text{cm}} + L_{\text{rel}}$

So I guess that $L_{\text{rel}} = [\frac{1}{2}\, u\dot{r}^2 - V(r)]$, explaining the form. Unfortunately, that means that L_{cm} is just a kinetic energy, with no potential energy component. Since when can you write a Lagrangian as just a straight kinetic energy? What possible least path is that action taking? It can't be a least or most path, since it can't vary. It is just one thing, and therefore cannot be pushed to into a least path.

Then we are told, "The **R** equation from the Euler-Lagrange system is simply Ma = 0 [where a is the acceleration of R, R dot dot], resulting in simple motion of the center of mass in a straight line at constant velocity." Well, I didn't need these equations or the Euler-Lagrange system to tell me that! Of course the center of mass is going to have an acceleration of zero, since you can't have a force there by definition. That is why we found a center of mass in the first place, for Pete's sake. This author at Wiki implies that he found the zero acceleration via these equations, but the zero acceleration was the postulate, so it cannot be the discovery. The mass causes the force, by definition, and the force causes the acceleration, by definition, therefore you cannot have acceleration at the center of mass (any more than you can have acceleration at the center of a single body). That is what center of mass means, by god.

But it gets even worse. We have already seen L_{cm} reduced to an idiotic tautology, now we also must see L_{rel} reduced to a rubble of finessed math. For the equation is then expanded via polar coordinates to this

$L_{rel} = \frac{1}{2} u(\dot{r}^2 + r^2\omega^2) - V(r)$

Where ω is the velocity or change in θ. Since L_{rel} is not dependent on θ, θ is an "ignorable" coordinate, we are told. It is ignorable, and there is "no dependence," which seems to be a great reason to find a partial derivative of L_{rel} with respect to it.

$\partial L_{rel}/\partial\omega = ur^2\omega = \text{constant} = \ell$

And of course ℓ is the conserved angular momentum. You have got to be kidding me. That's just pretend math, right? That equation was inserted as a joke by some mischievous math elf, right? No, apparently mathematicians and physicists really buy this stuff.

At least we can see why some of the previous equations were manufactured. We needed to get something we could differentiate into uvr or $ur^2\omega$. That is just the old angular momentum equation L = mvr. But it doesn't explain what happened to the potential energy, which just got washed down the drain. Since there is no angle in the potential energy, $V(t)$ just conveniently got jettisoned.

The only reason to take a partial derivative of L_{rel} with respect to that angular velocity is to push this equation, but there is no mechanical justification for it. First of all, differentiating requires a dependence. Remember first year calculus, where you were told what a function was? A function is a dependent variable, and in order to do calculus, you require dependent variables. Calculus requires functions, which requires dependence. But here they admit that the Lagrangian is not a function of the angle, then they differentiate the Lagrangian with respect to the angular velocity! Incredible chutzpah.

Beyond that, I have shown that angular momentum (see my first book) is not equal to mvr or $mr^2\omega$, which means all this equation finessing was in vain. Someone should have told them the historical angular momentum equations were false, so they could push these equations toward the right ones. As it is, it just makes it easier for me to see they were cheating. I can see that they were just pushing the equations toward what they thought they needed.

Now let us return to the Lagrangian for celestial mechanics. I have said that the Lagrangian is a poor or partial attempt to express the unified field. Since I have now written basic equations for the unified field, it might help to compare the Lagrangian directly to them. Let us use my force equation, to start with

$$F = (GmM/r^2) - (2GmM/rct)$$

That already looks a lot like the Lagrangian, doesn't it? Let us multiply both sides by r, as they do now, to make F into E. That gives us

$$E = (GmM/r) - (2GmM/ct)$$

Since that first term is just V, We could just rewrite that in Lagrangian variables, as

$$L = -V - T$$

My unified field appears to be a sort of negative Lagrangian. But does the kinetic energy equal 2GmM/ct? For a celestial body, I would have to show that ½ mv² = 2GmM/ct. Or that v² = 4GM/ct. Or that ar = 2GM/ct. Since a = GM/r², we get

GM/r = 2GM/ct
r = ct/2

Is r = ct/2? Well, if the unified field is mediated by photons, and it is, then yes, we would expect r = ct. The equation r = ct is just a sort of very simple relativity transform, and in fact both Lorentz and Einstein used it to develop the current transforms. We wouldn't expect the 2. The 2 means that my equation would actually translate into

L = -V – 2T

Which is more like a negative virial than a negative Lagrangian. But still, I have shown what I set out to show. The Virial and Lagrangian are variants of my unified field equations. The only difference is, I would never put an orbital velocity into a kinetic energy equation. That is what we had to do, as you see, to get the Virial from my unified field equation. But since we see the mainstream do stuff like this all the time, we know this is how the Virial and Lagrangian got into the sloppy form they are now in.

Let me clarify that. Velocity is a vector, so in the kinetic energy equation $T = ½ mv²$, it must be linear. But in the equation a = v²/r, the velocity is not linear. The velocity there is *orbital*. Therefore, that substitution I used was not really allowed. Even if you allow me to correct the equation, so that it is now a = v²/2r, the velocity is still not linear, and the substitution is still not allowed. In that correction, the velocity is still orbital, not tangential. To use a kinetic energy equation in the form ½ mv², we have to use a linear or tangential velocity. I have shown that the corrected equation for tangential velocity is v² = a² + 2ar, which would make the above substitution a² + 2ar = 4GM/ct, dooming the move from unified field to Virial or Lagrangian.

What does that mean? It means that although *T* looks sort of like a kinetic energy in the Virial and Lagrangian, it isn't. In both celestial mechanics and quantum mechanics, you have to force fit the current equation to make it work. In celestial mechanics, you have to pretend that you can put an orbital velocity into a kinetic energy equation, but you can't. *T* isn't really the kinetic energy, it is just a term that mimics the kinetic energy in form. The Virial and Lagrangian aren't really using the kinetic energy and potential energy, they are mirroring the terms of my unified field equation, and my terms aren't standing for kinetic energy and potential energy. They aren't even standing for charge and gravity. They are just two terms in the equation, unassignable directly to any real field or energy.

This means that if you use a real linear velocity in the Lagrangian, you are going to get the wrong answer. You actually have to use a bad or false expression for the kinetic energy to get the Lagrangian and Virial to work. You have to use a false substitution, of orbital velocity for tangential velocity, to make the Virial or Lagrangian work. This is what I meant when I said you had to push the Lagrangian in the right way, above. You have to insert the proper numbers,

76

which turn out to be fake kinetic energies, expressed with orbital or curved "velocities" instead of real linear or tangential velocities.

To see what I mean, we have to go back to the equations leading up to my unified field equation. These are taken from my first UFT paper.

$$F = E + H$$
$$F = (GmM/R^2)(1 - 2R/ct)$$
$$F = (GmM/R^2) - (2GmM/Rct)$$

E is the charge field and H is the solo gravity field. F is the unified field. But *neither* E nor H is expressed by GmM/R^2. That term is just Newton's equation, which was already unified. The other manipulation here is just my correction to Newton. That correction was found by segregating the two fields, then doing relativity transforms on both separately, then recombining them. So the term $2GmM/Rct$ is a correction, not a field. It is not the charge field, it is not potential, and it is not kinetic energy. But the Lagrangian is mimicking this equation, as I have shown. *V* is mimicking the first term, and *T* is mimicking the second, so that the Lagrangian is really this equation in disguise. *T* is not the kinetic energy, *T* is this correction to Newton.

And this has been another major problem with unification. Physicists since the time of Einstein have been trying to unify QM with gravity, but since the equations of QM are grounded by the Lagrangian, QM is already unified. Not realizing this, physicists try to unify their Hamiltonians, connecting them in various ways to GR (General Relativity). As I have just shown, this can only cause a mess. Everyone is trying to unify equations that were already unified. The reason they don't know this is that the Lagrangian was fudged, centuries ago—pushed to match data—and the push just accidentally matched fairly closely the unified field equation.

Yes, we can now see for certain that the equation finessing by Lagrange and Hamilton and the rest was completely accidental. We know that they didn't realize the equation was a UFT, because if they had known that, they wouldn't have later tried to unify it.

In conclusion, we have learned many things about the Lagrangian. One, the variables are misassigned. V is not potential, it is Newton's gravity equation, which was already unified from the beginning. And T is not kinetic energy. T is simply a term that corrects V, as in my unified field equation. T is a correction to Newton's F. It just happens to mimic the form of kinetic energy. But to make T work in the Lagrangian, you have to insert an orbital velocity for v. In other words, you have to insert a falsified kinetic energy. If you push the equation in the right way, you may get the right answer. But in most cases, the Lagrangian is just used as a fudge, as in the two-body problem.

In a subsequent paper, I will show that T also has to be pushed when the Lagrangian is used in quantum mechanics. If you use a real kinetic energy, rigorously defined, the Lagrangian and Hamiltonian fall apart. The only way to make them work is to use a fake "kinetic energy", one that has been pushed to match my unified field equation.

This means that we should dump the Lagrangian and use my

unified field equation instead. We should fix all these errors, so that we can see the mechanics and fields underneath our equations. If the form of my UFE is not what is needed for certain problems, it can be easily extended into other forms, some of which I have already provided. Using my UFE will allow us to solve many problems that have remained insoluble, at both the quantum and the celestial levels. In fact, I have already solved many of these problems in other papers. My UFE brings the charge field into the light, with all its mechanisms, and a hundred problems have already fallen to its clarity.

Chapter 6

An Analysis of
DARK MATTER

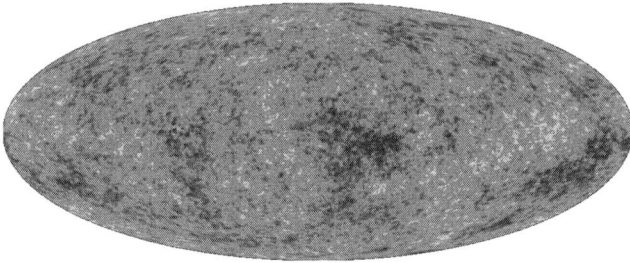

Let us look at the first paragraph at Wikipedia, under the title Dark Matter.

In astronomy and cosmology, dark matter is matter that is inferred to exist from gravitational effects on visible matter and background radiation, but is undetectable by emitted or scattered electromagnetic radiation. Its existence was hypothesized to account for discrepancies between measurements of the mass of galaxies, clusters of galaxies and the entire universe made through dynamical and general relativistic means, and measurements based on the mass of the visible "luminous" matter these objects contain: stars and the gas and dust of the interstellar and intergalactic medium. It is probably cold and if so, probably weakly interacting massive particles or many primordial intermediate mass black holes between 30 and 300,000 solar masses, or both.

The second sentence should read, "Its existence was hypothesized to account for discrepancies between *calculations* of the mass of galaxies. . . ." General Relativity is a math, not a measurement. The mass of the universe is a calculation, not a direct measurement. This means that the possibility exists that the math is wrong, and has been since the beginning of this mess. You should find it amazing that this possibility is so quickly dismissed, despite the fact that our math is known to be wrong in hundreds of other ways. Disregarding for the moment all the ways I have shown that mainstream math is compromised, the mainstream itself was forced to admit this a few years ago, when they reported a 15% general error in distance measurements. They downplayed the crushing importance of this, of course, but an error that size in something so basic is like a sky falling on modern theory. Again, these are distance *calculations*, not measurements. You can't measure astronomical distances directly, as with a yardstick. You have to use math to come to a distance *estimate*. Which means that the previous math was very wrong. If you correct not just the distance calculation, but also every bit of math that depends on distance (which would be just about all of it), you get total errors way over 100%. That is, your margin of error exceeds your data, so that your math is useless.

We can see that the margin of error has exceeded the data and the math in modern theory just by looking at the last sentence of the first paragraph. I would call that sentence a weakly interacting sentence, since it leaves the reader with absolutely no confidence the writer knows what he is talking about. We have two big squishy possibilities that aren't even remotely related to one another, and both possibilities come with zero data. They are wild speculation; and they aren't even what one would call good speculation, since they rely on stories made up just for the occasion. Meaning, they are completely *ad hoc*. Nothing exists to recommend either of those possibilities to the rational and reasonable, and they are on the table only because someone happened to think of them. That is like saying "I had a dream of a unicorn last night, and not of a griffon, therefore I think I will run with the unicorn theory for now."

After this thoroughly flaccid opening paragraph, we get shunted immediately into misdirection. We are told,

Ordinary matter accounts for only 4.6% of the mass-energy density of the observable universe, with the remainder being attributable to dark energy. From these figures, dark matter constitutes 83%, (23/(23+4.6)), of the matter in the universe, while ordinary matter makes up only 17%.

Unfortunately, that conflicts with the diagram posted:

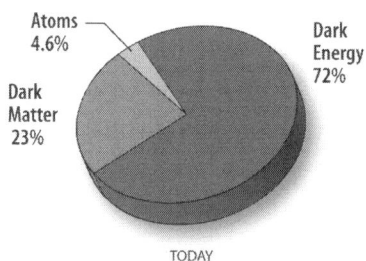

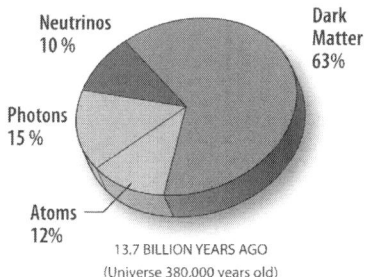

Atoms 4.6%
Dark Energy 72%
Dark Matter 23%

TODAY

Neutrinos 10%
Dark Matter 63%
Photons 15%
Atoms 12%

13.7 BILLION YEARS AGO
(Universe 380,000 years old)

83

As you see, you have been equation finessed at Wiki once again, to get the number down from 95.4% to 83%. They are trying to separate dark energy and dark matter, but matter and energy are interchangeable, according to Einstein and therefore current theory. Besides, they admit below that that dark matter is not thought to be baryonic or atomic. Therefore, we don't care what percent of "matter" is dark matter, since dark matter is not matter as we know it. Matter as we know it is baryons and leptons and so on. To remain honest, dark matter theorists should always lump dark matter and dark energy together, since this is what tells us how much missing mass/energy they have: over 95%. That is how much their calculations fail.

Actually, to remain honest, they shouldn't be allowed to separate dark energy and dark matter at all, ever, since they have not a spot of data that indicates a separation. The separation is completely theoretical, and the numbers 72 and 23 have been pulled out of a hat. Those numbers don't even come out of the equations that give them the 95.4% number. The numbers 72 and 23 don't come out of any equations at all. Those numbers, as well as the division into dark matter and dark energy, were created as damage control, and nothing more. We can see this immediately if we take the link to the dark energy page, where we read this:

Two proposed forms for dark energy are the cosmological constant, a constant energy density filling space homogeneously, and scalar fields such as quintessence or moduli, dynamic quantities whose energy density can vary in time and space.

You have to laugh. Seventy-two percent of the universe is composed of a mathematical constant. Of course that begs the question, "a constant assigned to what?" You can't fill mass deficits with a Greek letter. You also can't fill them with fancy words like *quintessence* or *moduli*. That just looks like a return to Aristotle. Modern physicists using the word *quintessence* is a lot like neocons choosing the title Homeland Security. It is either incredible chutzpah or incredible ignorance. Hitler used the term Homeland Security for his own Nazi government organization of intimidation, which should keep future generations off the term forever. Aristotle used the *quintessence* or *aether* as his fifth element, along with fire, earth, air and water, which should keep modern science from using it. There are many reasons for this, but I will hit only the big two: 1) they have been belittling anyone who so much as breathed the word *ether* for more than a century, ever since the Michelson/Morley interferometer is said to have disproved it, 2) they have been belittling metaphysicians for just as long, ever since the positivists are said to have destroyed metaphysics. For this reason, I find the return of the word *quintessence* doubly and triply strange. It is not so much that current physicists have decided to embrace metaphysics again, it is that they have embraced magic and irrationality without even realizing it. These new theories in physics don't even have the rigor of the old metaphysics, since they don't include any logic. New science has thrown out both physics and metaphysics, and is just subsisting now on bluster and fudge.

After that, we are told this,

Adding the cosmological constant to cosmology's standard FLRW metric leads to the Lambda-CDM model, which has been referred to as the "standard model" of cosmology because of its precise agreement with observations.

Its precise agreement with observations? You have to be kidding me. Do I have to explain this to anyone? A constant is a mathematical entity chosen to fill a hole in an equation. Therefore, the fact that it fits this hole is not a big surprise. Claiming that Lambda is a good constant because it fits observation is like saying G is a good constant in the equation $F=GMm/R^2$ because it fits observation. *Of course* Lambda fits observation, you idiots, since you chose it to fit observation. And when it failed to fit new observations, you changed it. You have changed it umpteen times to fit new observations, so using this good fit as proof of the theory is absurd. These new theories must be written only for the illiterate, since no one who knows how to read or who has taken highschool physics would fall for this stuff.

So we see that dark energy is a made-up term, made up to deflect you from noticing that this new dark theory has no explanation of 95.4% of the known universe. We also find out that even this high number has been pushed lower.

He [Fritz Zwicky, 1933] applied the virial theorem to the Coma cluster of galaxies and obtained evidence of unseen mass. Zwicky estimated the cluster's total mass based on the motions of galaxies near its edge and compared that estimate to one based on the number of galaxies and total brightness of the cluster. He found that there was about 400 times more estimated mass than was visually observable.

That is interesting for several reasons. One, 400 times is much greater than 19 times. 95.4% comes out to about 19 times. 400 times is about the same as 99.75%. So Zwicky would have needed 99.75% dark matter/energy to fill *his* miscalculation. Two, we are told that Zwicky used the virial theorem as his primary math. That is important because I have written several papers on the virial/Lagrangian, showing precisely where it is compromised. Yes, it is the field equations that are compromised, in not one but dozens of ways.

We see this again in the paragraph below that:

Much of the evidence for dark matter comes from the study of the motions of galaxies. Many of these appear to be fairly uniform, so by the virial theorem the total kinetic energy should be half the total gravitational binding energy of the galaxies.

The virial again, you see. I have shown that not only does the virial have an extra 2 in it, but the two main operators are misassigned. The virial/Lagrangian is a unified field equation, and always has been, and the operator T is misassigned. It is not the kinetic energy, it is only a term that mimics the kinetic energy in form. So when these physicists compare the total kinetic energy to the total gravitational binding energy, they are comparing terms that are not properly assigned. The terms are not what they think they are, so they are plugging the wrong numbers into the equation. This is why they get the wrong numbers out. I have proved this in great detail in a series of papers.

We are told that one of the odd properties of dark matter is that it doesn't carry any electric charge, and isn't affected by, or detected via, the electromagnetic field. Of course we could also say this of photons. Although photons are the quanta of electromagnetism in current theory, and although they are the cause of charge in my theory, in neither theory do the photons have charge themselves. That is, photons are not turned by E/M fields. Current theory doesn't tell you why this is, but they are well aware of the fact. In my theory, photons aren't turned by E/M fields because, individually, they are small enough to dodge the field. If the E/M field is defined as ions, photons dodge most ions easily. But the E/M field, at the foundational level, should be defined as other photons, and the photon field is mostly interpenetrable to itself, again due to the tiny size of the particles. The photon field is not completely interpenetrable to itself, of course. Nothing that is real is completely interpenetrable to itself. But the current model admits that the charge of the photon may have a real value below $10^{-35}e$. As it turns out, both the mass and the charge of the photon are above zero, and this non-zero "charge" is a measure of its density relative to itself. The charge of the photon is the amount each photon is affected by other photons.

I mention this fact because it ties into my own theory of missing mass. The missing mass isn't dark matter, it is the charge field. In other words, it is photons. Physicists have forgotten to include the charge field in their equations. Their gauge math tells them that photons have zero mass and zero charge, so the photons don't make it into the equations that way. And their total energy isn't properly included either, for much the same reason. The total energy of the light spectrum has been horribly underestimated, because the equations now used have failed to count up all the photons. This is because in order to estimate the number of photons in the total field, physicists now simply use macro-detections of the field. In other words, they measure the amount of incoming light at some point on the Earth, subtract out the Sun and solar reflections, and extrapolate up using various models. Or they use some other method equally naïve. But they totally ignore the charge field itself, since they seem to think this exists only at the quantum level. They make no attempt to measure the energy of the charge that has existed in their basic equations almost from the beginning. I will show you once again what I mean. Instead of estimating the total photon energy in the ways they have, they should have estimated it this way:

$e = 1.602 \times 10^{-19}$ C

$1C = 2 \times 10^{-7}$ kg/s (see definition of Ampere to find this number in the mainstream)

$e = 3.204 \times 10^{-26}$ kg/s

Those first two equations I took straight out of the old books. You can find the equations at Wikipedia. They aren't any inventions of mine. I simply combined them to get the third equation. The third equation doesn't look too revolutionary, until you remember that it means that the electron is emitting about 35,000 times its own mass every second, as charge. It also means the proton is emitting about 19 times its own mass every second. If we give this charge to real photons instead of to virtual photons, we have a simple way to estimate the total mass/energy of the photon field. It is 19 times the atomic field, or 95% of the total mass/energy of the universe.

Now, ask yourself this. Do you want to keep following a standard model that insults your intelligence by assigning 95% of the universe to unassigned constants, dark matter, WIMPS, or black holes; or do you want to switch over to a physics that treats you like a rational entity? I should think any person of good judgment would prefer to come over to my side, where we solve problems in three lines of simple math. As I hope you see, I found the number 95% in three or four lines of simple math. But if you like filling blackboards with Hamiltonians, talking about quintessence and moduli, and watching Hawking and Penrose debate about the precise location of a wormhole, then stick with the standard model. You might also want to get a tall pointy hat with stars and moons on it.

Chapter 7

How my Unified Field solves the

GALACTIC ROTATION PROBLEM

and how the dark matter math is fudged

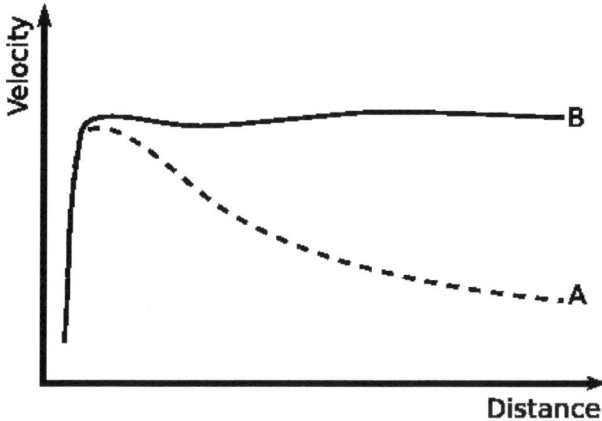

In a long paper on the Allais Effect [allais.html], I mentioned several modified Newtonian fields of the past, including those of Quirino Majorana and others. Once we are familiar with these older (and failed) modified Newtonian fields, we can see the newest modified field in the same light. I am speaking of course of MOND, the Modified Newtonian Dynamics of Mordehai Milgrom [from about 1983]. While this MOND is preferable to the theory of dark matter, and while it does move slowly in the right direction, it is woefully incomplete (as I think even Milgrom would admit). Lee Smolin has put it this way:

MOND is a tantalizing mystery, but not one that can be solved now.

He means by that that MOND seems to point to some new field, but that it is not a field we know about. Smolin is both right and wrong. MOND *is* tantalizing, in that it implies a new field; but the mystery has already been solved: the "new field" is simply the charge field. MOND is pointing directly at my unified field.

The problem is that Milgrom's new function μ is physically and mechanically unassigned. It is basically just a fudge factor or a hole filler, a piece of raw math with no theoretical or field underpinning. In other words, we are never told what causes it.

The form of Milgrom's math is also a clear problem, since, like Majorana and the others before him, he adds the field outside of Newton's equation. In MOND, the rotational velocity is found by this equation

$$v = \sqrt[4]{GMa_0}$$

As you see, his constant is separate from G and M.

But I have shown that Newton's equation is correct, as written. This is why I do not call my unified field a modification of Newton. I have not modified Newton's equation, I have pulled it apart and interpreted it more fully.

You see, the trick is that the neceassary variations are already contained within G and M, so we do not need any new functions or constants. We just have to understand what G and M really stand for, and how they work mechanically.

This is why the solution has been so difficult to see. It was hidden in the constants and variables we already had.

The reason I was able to solve this galactic rotation problem so fast is that I had already solved it before I even knew of it. The solution has been sitting in plain sight in my other papers, waiting for me to become aware of the MOND problem. You will be surprised that I have remained ignorant of it until now, but my critics are not completely wrong when they point out my isolation. Being self-taught and non-affiliated, I have been making my own way, at my own rate. This situation has both plusses and minuses. The main plus is that I am free to think what I like, and go where my mind leads me. The minus is that I am sometimes late coming to a problem, as here. I read of the problem last night, and solved it this morning.

Let me gloss my unified field one more time. The Universal Gravitational Constant G is the key, since it is not a naked constant, but a scaling constant between two fields. Newton's equation, as written, already contains the charge field, and G scales between charge and gravity. All we have to do is write each mass in the equation as a density and a volume, giving the volume to the gravity field and the density to the charge field. G then scales between them, taking the atomic size down to the photon size. This is necessary because gravity is relationship between atoms, or things made of atoms, while charge is relationship between photons and atoms. For charge to work, photons have to collide with atoms or ions, and this requires a scaling between the two particle sizes. I explain this much more fully elsewhere.

Once Newton's equation is interpreted in this way, we find subtle changes in the field. I have already uncovered many of these changes in other papers [see especially the two-mile problem], but in this case the change becomes quite large and obvious. The velocity divergence in outer arms of galaxies is very large, and is not what anyone would call subtle. This is why I found this problem so interesting, and why I had to leap on it instantly. It is solved immediately not by tweaking the equation, but by recognizing the variation in the field. Put simply, the charge field variation from center of galaxy to outer reaches is very large. Since all matter emits charge, there will be much more charge near the center of the galaxy.

The mainstream has missed this obvious field variation for many reasons. One, they give charge to the messenger photon, which is a virtual photon. Since virtual photons do not take up any space in the field, they could not provide any drag. Two, even if they began giving the charge field to real photons, their real photons also do not take up any space in the field. Since they are point particles in the math, they cannot take up space or provide drag in the field. Three, all mainstream theories have ignored the charge field completely. They have tried to solve this problem with gravity alone, or gravity plus relativity, or gravity plus unassigned constants. They have not seen that Newton's equation must include the charge field. Nor have they seen that if Newton's equation includes the charge field, it must cause both subtle and unsubtle variations in the field mechanics.

I will be told that my solution requires more than just a re-interpretation of Newton's equation. It also requires a re-interpretation of the photon. Newton's equation, by itself, has nothing to say about the photon. True enough. However, it may be worth pointing out that Newton did not think the photon to be virtual or to be a point. He agreed with me since he agreed with Descartes on this question: anything that exists has extension. The photon exists, therefore it must have a radius. If it has a radius, it takes up space. If it takes up space, it must provide drag.

The mainstream should know this, since they admit that the photon has momentum and energy. A thing with energy and momentum could hardly be "invisible" in the field. It could not act as a ghost, regarding drag. A photon could not be capable of knocking electrons out of atoms, but incapable of taking up space in the field. A mathematical point cannot knock an electron out of an atom. This is because scattering could be thought of as a kind of drag. Both would be caused by real collisions. You cannot propose that the same particle can cause scattering, but cannot cause drag.

The reason Milgrom's MOND was so tantalizing is that he had the right variation, in a way, and also the right explanation of it. He said that his function did not cause measurable variation in the solar system because gravity is so strong here, near the Sun. That is not precisely correct, but it is a good hint. It is not because gravity is so strong here, it is because the Sun is the only major body acting as the central mass, making the velocity follow the inverse of the radius (see below).

This is why the solar system seems to follow gravity, while the galaxy seems to follow charge. The unified field relates gravity to charge, and this makes Milgrom nearly correct. The mass per volume in the solar system acts to flatten out the charge field variations due to the sphere, causing the bodies to follow classical equations pretty closely. This is just one more reason charge has been invisible to us. The local charge field variations are swamped by gravity, and so we have ignored them. Only when we look at the galaxy do we see how large these variations normally are.

This is also why globular clusters show much less charge variation, and therefore much more velocity variance, than galaxies. It is simply a matter of density distribution. Globular clusters are known to have much higher star densities than galaxies, and, as with the solar system, this density tends to flatten out the charge variation. Less charge variation means greater velocity variation, as we will see in the equations below.

Now, let us look at the dark matter hypothesis for a moment. Initially, it was said that around 50% of the matter of a galaxy must be in a galactic halo, completely outside the visible galaxy. This number is now about 95%. Yes, a big problem required a big solution, and this tells us how large the velocity variance from prediction really was. If you visit a place like Wikipedia, you find dark matter proposed as the solution to velocity variance, but you get no math or theory. How does dark matter in the halo, even at 95%, cause a flat velocity? If the answer were clear, you would think Wikipedia would take the time to gloss it. It shouldn't take long, should it? This is a big clue. Wiki usually likes to cloak the theories with math, but here we get nothing. That must mean the math is really pathetic.

You only have to do the baldest math to see that extra mass out there can't solve the problem. Remember, the mainstream math doesn't have my charge field, so they are not using charge to flatten the field as I did, as in the solar system. They are adding mass but not charge. If you simply add mass to a halo, you cannot increase an orbital velocity. This is because orbital velocity has nothing to do with mass. It has to do only with distance from center. $a = v^2/r$, remember? No mass variable there. The dark matter hypothesis is not any type of MOND, so it is not modifying Newton at all. Well, according to Newton's equations, the orbital velocity is determined by distance *and nothing else*. Jupiter's orbital velocity is not determined by its mass, and if we took the Earth out to the distance of Jupiter, it should have the same orbital velocity as Jupiter.

The dark matter hypothesis needs to tell us more than "dark matter." It needs to tell us how dark matter can cause greater orbital velocities. Say there *is* a huge amount of dark matter in a halo around most galaxies. How will this affect the velocity of shining matter that we can measure? Well, it would tend to suck all the matter out of the galaxy toward the halo, but it wouldn't affect the orbital velocity of that matter at all (unless it decreased it—see below). This is because, according to both Newton and Einstein, gravity has no force at the tangent. A gravity field cannot apply a tangential force, only a centripetal force. Therefore, a large halo could only pull out on matter inside it. It could not pull sideways. According to both Newton and Einstein, gravity can neither cause nor increase an orbital velocity. It has no mechanism to do so. According to the explicit math of Newton, an orbital velocity is the compound of a centripetal force of gravity and an innate motion of the orbiter—this innate motion being the tangential or straight-line velocity. Since gravity and the innate motion are independent, gravity can never affect the innate motion. Therefore gravity cannot cause the orbital velocity, much less change it. It can only determine the radius, given the innate velocity.

The mainstream* try to get around this by using this equation

$$v^2 = MG/r$$

Whereby, if you increase the mass you increase the velocity. Unfortunately, that is gigantic fudge, since M is the mass of the center, not the mass of some body in orbit. That equation comes from solving these three equations

$$F = ma$$
$$F = GMm/r^2$$
$$a = v^2/r$$

M is explicitly defined as the large central mass causing the field, not the mass in orbit. In the case of a galaxy, it would be the mass of the galactic core, not the mass of the galaxy as a whole or of a body orbiting the galactic core. Therefore, their math is completely upside down. They have "solved" only by ignoring the explict definitions of the variables in the equations. At the Duke University website linked below, the author states

M = mass lying *within* stellar orbit

That proves my point, since that means we are talking about the galactic core here. Since the core is not a discrete object in a galaxy, all the mass inside the radius can be called the core. M is the central mass, not the orbiting mass. This is crucial, since the mass and the velocity both increase as r increases. An increase in M doesn't just increase velocity in the outer reaches, it increases velocity everywhere proportionally, like an increase of a.

This means that the dark matter math is also a type of modified Newtonian dynamics. It is modified in that it takes the definitions and turns them upside down. We could call it a FFAND: a falsified and fudged Newtonian dynamics.

What this equation of Newton $v^2 = MG/r$ is actually telling us is that if we increase the mass of the galactic core, we can increase *all* the orbital velocities, at all radii. But it is not telling us that we can flatten the field in any way. And if we increase the mass of a halo, we have thereby relatively decreased the mass of the core, which must DECREASE all orbital velocities. **Dark matter would make the problem worse**.

Another major problem with the dark matter solution here is that a halo with that amount of matter could not remain undetectable in our own galaxy. We always hear of dark matter supposedly found here and there, as WIMPs or whatnot, but of course the best place to look would be in our own halo, would it not? If 95% of the mass of the galaxy is in a halo, then it cannot be invisible to all detection, dark or not. Remember, we are in an outer arm of our galaxy, and therefore we would be quite near this halo. Unless we are looking toward the core, we are looking through this halo whenever we aim our telescopes into space. With a mass nealy equal to the entire galaxy, this halo must have an appreciable density. Why doesn't it affect our extra-galactic views? We can't detect it even indirectly, as it shifts or distorts or tamps down incoming data. It doesn't really matter if it is baryonic or non-baryonic: if it is real, it must have density, no matter what it is made of. It cannot have mass and lack density, can it? A thing can have mass and lack density only by changing the definition of mass or density. That is what all the larking about with non-baryonic matter is: the attempt to convince you, by some sort of speechifying, that matter can have mass but no density. WIMPs and axions and hidden sector particles are all "heavy" particles with no density. They are heavy ghost particles, in other words. If you can imagine heavy ghosts, good luck to you. The fact is, "heavy" and "weakly interacting" are mutually exclusive adjectives. A thing can interact weakly only if it is small or diffuse, and if it is small or diffuse, it cannot be heavy. Once more, this is just the sad attempt to change the definitions of words. As the void is now a thing, and as virtual particles now cause motions, we have heavy massive particles that are both weakly interacting and undetectable. It is so pathetic it truly defies belief.

I believe in some dark matter. The earth is a bit of dark matter, of course. But these theories of weakly interacting massive particles could not be more silly. The fact that anyone takes them seriously is a sign of the nadir. We don't need WIMPs, we need to understand that photons have mass and radius. All these ridiculous problems and theories are caused by refusing to let the photon be a real particle.

Remember that I have shown that charge is equivalent to mass. But the standard model has not gotten that message. They have forgotten to weigh all the charge in the universe and to include it in their energy totals. They don't even include the weight of the E/M spectrum in their totals, telling us that photons have no rest mass. In other words, they don't include the mass of visible photons, much less the mass of invisible photons that we already know about, like infrared photons and so on. The few models that do include normal photons in the mass of the universe only include a small fraction of them. For instance, I have shown that every electron is emitting a charge each second that outweighs it by 35,000 times, (see math below). All that charge is unweighed by the standard model, and is not included in their totals. That is why they need dark matter and dark energy in the amount of 95%. Dark energy is mostly charge. It is charge photons. This charge is also what is causing drag in the inner parts of the galaxy, creating a flat velocity line.

With this under out belts, we can return to the MOND equation for velocity.

$$v = {}^4\sqrt{(GMa_0)}$$

Once again the problem is with the mass variable M. Milgrom created MOND to compete with the dark matter math, but he accepted their definition of M. The fudge I uncovered at Duke has been embedded in the galactic rotation problem almost from the beginning. Milgrom does not use a capital M here by accident. He just took their math and varied it, using his new functions and constants, as is clear from this equation. Therefore, he has inherited their fudge. His mass is misdefined, so that his equation cannot prove what he hopes to prove. Because his mass M is the mass of the central body (galactic core, in this case), his equation is flawed at the ground level. An increase in that mass will increase the velocity, yes, but it will not flatten the graph. It will increase all velocities proportionally. To do what he wants it to do, the mass would have to be re-assigned, as with the dark matter math. But you cannot do that without cheating. To do that would not be a modification of Newton, it would be a complete refutation of Newton and his variables.

The form of Milgrom's equation also makes it impossible for him to solve this problem. As you will see from my math, what we need is a differential, but Milgrom's velocity equation is too simple. Both G and a_0 are constants, so that v can be a constant only if M is. But M cannot be a constant, as we saw with the dark matter math. M is the core, which in a galaxy is anything below radius r. This means that as we increase r, we increase M. So M is changing in Milgrom's equation. This means that v will also change, and the velocity is not flat.

On the other hand, if Milgrom defines M as the mass of a constant core, he is guilty of another cheat. He has rigged his equations so that the radius cancels, which means we don't know the radius for the velocity he is finding. He simply states that the equation is good for all radii, but his derivation doesn't show that. To the contrary, his equation requires a radius, and it is saved in the current form (to a small degree) only because mass will vary as radius does. If he redefines the mass as a constant, however, he has just contradicted his own derivation.

So let me now correct all this bad math and theory. As a first question, we may ask how dense the matter field, and therefore the charge field, would have to be in order to begin causing photon drag. Well, we know that the charge field is dense enough in the solar system to cause axial tilts and variations from Bode's law and perturbations and torques and magnetospheres and so on, so the charge field here is already dense enough to cause drag. All charge field phenomena could be called drags of one sort or another, and if the charge field can cause perturbations it can cause velocity variances. It is not lack of a charge field in the solar system that causes the planets' velocities to follow the inverse of the radius, it is something else entirely, as I show below. Therefore, a matter density such as we find in the vicinity of Neptune is more than enough to create the required photon density. If it were not, then the axis of Neptune could not be turned by the charge field.

But again, how dense is the charge field? I have shown that the electron is emitting about 35,000 times its own mass every second as charge. You will say, "Hold on there! I won't bother taking that link, since that is ridiculous." But I took it right out of current definitions:

$e = 1.602 \times 10^{-19}$ C
$1C = 2 \times 10^{-7}$ kg/s (see definition of Ampere to find this number in the mainstream)
$e = 3.204 \times 10^{-26}$ kg/s

That's 19 protons per second. If the charge photon has an average mass of around 2.77 x 10^{-37}kg, then that is around 1.15 x 10^{11} photons per second. 11.5 billion photons per second, by each charged baryon. Which is a density of .03 kg/m^3/s. We will make good use of that density in later papers.

So why don't we measure the charge field when we weigh things? Because the charge field is completely uncontained and cannot be weighed. It is traveling c in all directions, and has no rest mass. Despite that, its mass must be included in all totals. If standard model totals are correct, and 95% of the total mass is unaccounted for, then it would appear that photons outweigh everything else by about 19 to 1. After doing the math above, that is not hard to believe at all. In fact, the math just above generates the number 19. That is why the mainstream is getting a figure of 95%: 95% is the same as 19 to 1. Current physicists have the right number but the wrong explanation. It is the charge field that outweighs baryonic matter by 19 to 1, not dark matter.

All that was an interesting diversion, but we don't need to count up photons or weigh them in order to solve this problem. We can take some shortcuts, the biggest shortcut being G. We know that if the charge field drag is ignored or if it is constant, a spherical field can be simplified to $v = \sqrt{(ar)}$. But let's rewrite that to get a mass in it

$F = Gmm_0/r^2$
$F = ma$
$a = Gm_0/r^2$
$v = \sqrt{(Gm_0/r)}$

That mass is the central mass, or the mass inside the radius r. If we let m_0 be the mass of the entire galaxy and r equal the radius of the entire galaxy, that equation gives us a velocity of about 390 km/s for stars at the edge our own galaxy, which is close to the current value of 220 km/s. But this remaining difference indicates appreciable charge field drag even at the outer edge of the galaxy. It also indicates that the current numbers are wrong, since we don't have enough mass in the outer reaches to make up that much difference.

But now we have to include the charge field drag, to create a differential equation. In the above equation, we have the charge field included in the scaler G, so that the equation is already a unified field equation, but we have not indicated a charge presence in the field as size, so that the photons can create drag. As written, the equation only indicates the energy of the charge field relative to the gravity field, allowing the charge field to collide with matter and create the E/M field. But the equation does not include the separate but related ability of the charge field to create resistance or drag. To do that, we have to create a separate term in the equation, and subtract it from the first term. Like this:

$$v = \sqrt{[(Gm_0/r) - (GM_r/r)]}$$

This second mass is defined as the mass at radius r, rather than the mass inside radius r. This solves the problem of previous maths, which did not include both variables. This second term represents the density of the charge field at a given radius and allows us subtract it out as a sort of drag. Because the mass at that radius is multiplied by G, it becomes the emitted charge field instead of the matter field. In the first term, G scales between two field, both fields being represented in the term. But in the second term, G is simply taking the matter field and turning it into the charge field. In this way, the second term is able to represent the drag of that field. Many would have tried to solve by creating a drag equation, but this is a much simpler method of solving, as you can see.

Once we study the equation, it becomes clear why it gives us different slopes for the galaxy and for the solar system. This equation is actually the correct one for all systems, but in the solar system we approximate by ignoring the second term. If you insert some numbers, you find that the reason it doesn't create a flat line in the solar system is that the mass inside r is always about the same. With only small variations, the mass inside r is just the mass of the Sun. So m_0 doesn't change with different values of r, and this makes v change with r inversely. But in the galaxy, m_0 changes greatly with different values of r. All the mass inside r counts as the core, so it increases substantially as r increases. And as the first term gets larger, the second does too, which means the differential tends to remain nearly constant due to the density distribution of spiral galaxies.

Some will say that this new equation can't be right, since it gives us too large a variance in the second term for planets in the solar system. And if we apply the equation to the orbit of the Moon about the Earth, the variance becomes even larger. Am I really offering this equation as a general equation? Yes, I am, since these problems are easy to solve. First of all, the variance isn't that great, due to the square root, even with the Moon. And we also have other factors we are ignoring. Remember, in the solar system and Moon system, we have a charge field inside a greater charge field. In the case of the

Moon, for instance, the equation would be existing inside the much greater equation of the Sun's field. The Sun's charge field is much greater than that of the Earth, so it tends to tamp down the charge variations between the Earth and Moon. This also applies to the solar system, since the solar system is not only in its own charge field, it is in the greater charge field of the galaxy. Nonetheless, this new equation will help us fine tune all the velocities in all orbits. It will also force us to recognize the field presence of the photon, not only as charge but as resistance. This is the correct equation, and always has been. Historical and current equations are only attempts to derive this full unified field equation.

Yes, this is my relativistic unified field equation, in its velocity form. In an earlier paper, I developed the relativistic unified field equation, as a force, by a completely different method. Here, I developed the velocity equation from first postulates again, not using my UFT force equation. Fortunately, the two equations match, confirming both papers and both equations. You may study an even more recent paper (on my website) to show how the two equations resolve.

Other critics will point out that we have done experiments showing that photons coming to us from long distances are not affected by any ether, field, or "foaminess" of space. NASA recently published a video showing just this, in a long anticipated experiment. Shouldn't this disprove my equation and my theory? No, since the photon field is not affecting photons in this paper. The photon field is affecting matter here. I am proposing that photons have drag on matter, not that they have drag on other photons. I have never proposed that the charge field affects the linear speed of photons, or that it would affect small wavelengths more than large wavelengths. I have shown that it would change wavelengths, but not that it would change some more than others. Therefore, the NASA experiment and other experiments have nothing to say here.

From all this, we see that the problem has been that contemporary physicists do not understand Newton's gravity field. They don't even comprehend the variable assignments, and nothing is more basic. I have shown that this applies to both sides of this argument. It also applies to the non-symmetric gravitational theory of John Moffat, since Moffat just tries to hide behind tensors, and the conformal gravity of Philip Mannheim, who hides behind Riemannian curves. We do not need curved math or tensors to solve this. We just need to understand the variables and constants in Newton's equation.

Conclusion: we do not have to propose any modification to Newton or Einstein to solve the galactic rotation problem. Nor do we need dark matter. We simply have to recognize the charge field, which already resides inside Newton's equation. Once we do this, the problem evaporates.

*http://www.phy.duke.edu/courses/055/syllabus/lecture24.pdf, p.4

Chapter 8

A CRITIQUE *of the* CURRENT
BULLET CLUSTER
INTERPRETATION

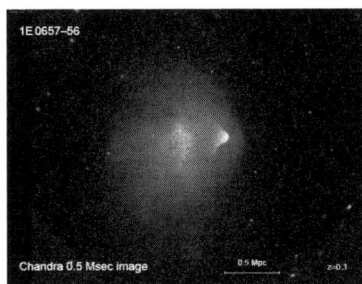

In 2004, Douglas Clowe et al published an 8-page paper* on the bullet cluster at IOP claiming to have proven beyond any doubt the presence of dark matter in the cluster, and thereby the universe. Soon after they published a 5-page letter [A DIRECT EMPIRICAL PROOF OF THE EXISTENCE OF DARK MATTER]** compressing and extending the earlier paper. Since then these papers have been used to assert the death of all MOND's and MOG's (modified Newtonian dynamics or gravities) and to trumpet the victory of dark matter. Even the opposition has been deflated by this paper, and many or most MOND supporters have backtracked, trying to save face by presenting new papers that show a mixture of dark matter and modifications. This in itself is a sign of the times, since it proves that neither side is capable of close analysis. The Clowe paper is extremely weak, as I will show below, and the fact that anyone would be cowed by it is a very bad sign for physics.

As those who have been keeping up with my papers know, I am not a supporter of MOND or MOG. I have proved, with simple equations, that both MOND and dark matter are false. So I am not here as a defender of MOND. I can show you where I stand very quickly by analyzing the first two sentences of the second paper:

We have known since 1937 that the gravitational potentials of galaxy clusters are too deep to be caused by the detected baryonic mass and a Newtonian r−2 gravitational force law (Zwicky 1937). Proposed solutions either invoke dominant quantities of non-luminous "dark matter" (Oort 1932) or alterations to either the gravitational force law (Bekenstein 2004; Brownstein & Moffat 2006) or the particles' dynamical response to it (Milgrom 1983).

Both the first sentence of the paper and the second sentence of the paper are false, which does not bode well for the rest of it. The first sentence is true only if we do not include the charge field emitted by the "detected baryonic matter." But if we allow that all baryonic mass emits a charge field, then this charge field must be included with any detection of matter. The standard model admits that all mass includes a charge field, so the sentence is false. To make the sentence true, we would have to make it,

Provided that we refuse to give the charge field real mass or mass

110

equivalence, we have known since 1937 that the gravitational potentials of galaxy clusters are too deep to be caused by the detected baryonic mass and a Newtonian r−2 gravitational force law.

The second sentence is also false, since what its should say is that the only *widely publicized* solutions are of those two sorts. As with the two political parties, Clowe wants you to believe you only have two choices. But since my solution invokes neither dark matter nor new functions, this sentence must be false. My solution exists, therefore it has "been proposed", therefore sentence two is false.

But we get misdirection even before we get to the body of the paper. The first sentence of the abstract is this:

We present new weak lensing observations of 1E0657−558 (z = 0.296), a unique cluster merger, that enable a direct detection of dark matter, independent of assumptions regarding the nature of the gravitational force law.

The authors are claiming a direct detection of dark matter, which means they don't know the definition of "direct detection." At best, with loads of assumptions and interpretations, they may claim an *indirect* detection of something unexplainable, which they then call dark matter. But there is nothing even approaching a direct detection of anything here. Then they claim that this direct detection, which is not a direct detection, is independent of assumptions regarding gravity. What they should say is that their paper is not a MOND, which should be fairly obvious considering the title. But their assumptions regarding gravity are omnipresent. To start with, they have just said that they are using weak lensing observations, so they must be assuming that gravity causes lensing. Beyond that, they are assuming the charge field doesn't have mass, that the photon is virtual, that gravity is absent at the quantum level, and a host of other things. This means that their assumptions are all mainstream assumptions, as of 2006, but it does not mean that their theory is independent of assumptions.

The last sentence of the abstract is also strange:

An 8σ significance spatial offset of the center of the total mass from the center of the baryonic mass peaks cannot be explained with an alteration of the gravitational force law, and thus proves that the majority of the matter in the system is unseen.

First of all, the paper doesn't show a spatial offset of the center of mass from the baryonic center of mass, it shows an offset from the plasma. The baryonic center of mass is roughly taken as the plasma center of mass only by claiming that the stellar matter is only 1-2% of the total, and therefore around 10% of the plasma total. Therefore the plasma stands for the baryonic center of mass here, within 10%. But the authors never prove that the stellar matter is 1-2% of the total. All they do is make one citation. Then they say this proves that the majority of the matter in the system is unseen. But even if all their assumptions and citations are 100% true and proved, this last proof of "unseen matter" does not prove dark matter. Why? Because I also believe that the majority of the mass in the system is unseen, but I do not believe in their dark matter. Notice that if I am correct, and if this "unseen matter" is my charge field, it would confirm the last sentence of their abstract. The charge field is dark and outweighs the baryonic field by 19 times (see below and my other papers). Therefore the majority of the matter in the system is unseen. The authors can prove their thesis regarding unseen matter, and still fail to justify their

112

title.

And that brings us to the title, which is even more bold than the abstract:

A DIRECT EMPIRICAL PROOF OF THE EXISTENCE OF DARK MATTER

We actually have neither a direct detection nor a direct empirical proof. To be clear, a direct empirical proof would be something like a unicorn in a box, as a proof of unicorns. What this paper offers is nothing even remotely resembling a unicorn in a box. What it offers is an extraordinary display of fudging, massaging, and finessing of equations and data (and the English language) to achieve a propaganda coup, one that will no doubt win someone a Nobel Prize in a few years. The paper is written expressly to impress the sort of spongy careerists who sit on these committees, so the future is almost predictable.

Yes, this paper of Clowe et al is another piece of unsubtle political posturing. It is attempting to hypnotize you from the first word, to convince you that "we have known since 1937" that this is not solvable with normal matter, and that the solution must be one of two. Only if you happen to be one of the multitudes of contemporary physicists who don't know what words mean—who don't know the difference between "direct" and "indirect" for example—will their paper convince you.

In paragraph 2 of the introduction, we are told this:

However, during a merger of two clusters, galaxies behave as collisionless particles, while the fluid-like X-ray emitting intracluster

113

plasma experiences ram pressure. Therefore, in the course of a cluster collision, galaxies spatially decouple from the plasma.

That is the first major unproved assumption here, unclothed and standing in full daylight. In fact, it would be true only if the dark matter hypothesis were true, so the argument has already gone circular. The point of the paper is to show that the data proves the dark matter hypothesis, but we have an unproved dark matter assumption here in the second paragraph. Circular. Galaxies *won't* behave as collisionless particles unless they are mainly composed of so-called non-baryonic matter, in which case they behave as ghosts. But if galaxies are mainly composed of normal matter, they won't act as collisionless particles. They will act as fields with real densities. Therefore, if it is proven that plasma decouples from matter, it must be because of density variations, not the fact that galaxies are mainly collisionless. It is known that plasmas do have different densities than normal matter, so data showing decoupling is not proof of dark matter.

In paragraph 4, we find this:

In the absence of dark matter, the gravitational potential will trace the dominant visible matter component, which is the X-ray plasma. If, on the other hand, the mass is indeed dominated by collisionless dark matter, the potential will trace the distribution of that component, which is expected to be spatially coincident with the collisionless galaxies.

This is nothing but (semi)clever misdirection. The authors tell us that the X-ray plasma is the dominant visible matter component, but that has never been proven. Just because some physicists published a paper on it does not mean it has been proven. These authors treat a citation as proof. We are told that the "stellar component" of a galaxy is 1-2%

114

[Kochanek et al, 2003], while the plasma component is 5-15% [Vikhlinin et all, 2006], but those percentages are only based on models. What models? Dark matter models! So these authors are using as evidence for their dark matter model percentages from previous papers that also used dark matter models. Again, circular. Before dark matter came along as a theory, it was thought that a large percentage of galactic mass was stellar. We can only get the percentage down to 1-2% by assuming undetectable WIMPs of some sort, so that the bulk of the mass is near-zero density and near-zero interacting. Again, these authors are assuming what they are trying to prove.

The only true assumption here is that the gravitational potential will trace the most massive component of the galaxies. But if the dominant matter component is not the X-ray plasma, then neither in the presence of dark matter nor the absence of it will the potential trace it. In which case, we have no "on the other hand." Both the classical and the dark matter hypothesis would predict the same thing, and tracing the potential will not help us. These authors have again created two possibilities, making you think they are the only two. The paper rides on these stacked assumptions, and they have already become wobbly by paragraph 4.

We can see this just by looking at the percentage they cite for plasma of around 10%. Well, if the plasma is 10% of the mass, it can't be dominant, can it? They say that in the absence of dark matter, the potential will trace the plasma, but that is just a strawman. It isn't true. In models that lack dark matter, such as mine, the potential would *not* be expected to trace the plasma. This is because no logical model would give the majority of mass to the plasma.

Plasma has to be created by the normal matter, via the charge field, so logically it cannot outweigh it. Plasma has to be ionized, and ionization takes energy. You cannot propose a total transfer of energy from a less energetic field to a more energetic one. Therefore, plasma must be energized by a more massive field—the matter field.

We see this again in the authors' unstated method of figuring plasma mass. The logical way to do it is this: from the luminosity and X-ray temperature you get an electron density; from the electron density you get a density of hydrogen and helium (and the other elements if you like). But then you must multiply by a volume. The authors are assuming the plasma fills the whole volume, but this assumption is false. The plasma is much more likely to be in the form of an envelope or shell, so that the mass estimate would fall dramatically. In addition, the authors are assuming that X-ray luminosity traces the mass. This may be true in the optical but it is not true in X-rays. The luminosity traces the electron density and temperature only. There could be a huge mass of gas in the outskirts beyond what you see in the X-ray image, if the temperature of the gas is too cool to be detected by Chandra. So their X-ray centroids are centroids of X-ray *brightness* only. Therefore they cannot logically claim that these brightness centroids are mass centroids. In fact, the X-ray luminosity goes as electron density *squared* so you only need to drop the density a factor of 10 for the X-ray emission to drop below your instrument detection threshold. The drop in temperature will multiply this effect. But it is even worse than that, since the authors use an X-ray temperature/mass relation derived from nicely behaved single clusters which are in equilibrium (which still assumes the whole volume is filled). The bullet cluster is as far away from equilibrium as you can get for these systems. All these assumptions lie hidden beneath the threshold of the visible paper.

Correcting these mistakes may either increase or decrease the total mass of the plasma, depending on the losses and gains. No doubt the authors will rush to coopt these comments, using only the ones that could cause a gain; but the point is that the authors don't know how to do the fundamental physics here. Which gives a reader little confidence in their claims to have proven anything.

Another major misdirection is dividing "normal matter" into stellar matter and plasma, especially when you remember that stars *are* plasma, according to the current model. The dark matter theorists want you to think they are different here, but they are different only in that they have been decoupled due to different densities. In fact, they are both baryonic and both include the charge field, so we don't have three categories, we only have two. We have matter—which includes plasma—and we have charge. But these authors are using a divide-and-conquer method here, assuring you continue to look toward dark matter and away from charge. They want you to think we need new matter to fill a gap, and if you remember that stellar matter and plasma are basically the same thing, you might remember the charge field, which ionizes both.

But these problems are just the beginning. We haven't even made it to part 2 (Methodology and Data). There we are shown a map of the potential and told how it is generated. It is generated by assuming that gravitational lensing has been proved, and then applying the formula for weak gravitational

117

lensing to the raw data in order to stretch the image. Problem here is that gravitational lensing is just a mathematical theory, with a few poor data examples. In a long paper, I have shown that we have much stronger data against gravitational lensing than we have for it. The examples the mainstream always lead with, those of the Twin Quasar and Einstein's Cross, are actually very speculative, not to say illogical, and they don't come near to proving the theory. Since the two strongest candidates are so weak, we don't even need to look at weaker candidates. And the data *against* lensing is ubiquitous, since the dark sky we see every night is that data. If the theory of lensing were true, every object in the sky would be haloed and rehaloed to infinity. All the lenses would act as scattering, and we wouldn't have a dark spot in the sky.

Therefore, we may say, with no threat of contradiction, that the authors are using an unproved assumption to generate an image. Lensing is a hypothesis with absolutely no strong data to back it up.

But even if the lensing theory is true, there is no indication it would create a reduced shear across the *interior* of massive collections of objects. The original idea was that concentrated masses or collections of mass would create bends *around* them, and the idea that they would create bends on light passing *through* them was only tacked on later. For this idea, we have even less evidence. One might say that the data is zero and must remain zero, since to accept the idea, we have to accept that the image we see is not correct.

Just think about it for a moment. How would you prove that

one way or the other? What these theorists have done is simply *assume* that if light is bent by going around large collections of mass, it must also be bent going through them. Therefore it distorts them. Therefore all images are distorted. But in that case, all data can be used as proof for and no data can be used as proof against. Let me ask you this: Could you know if all images were NOT distorted by lensing? No, once you assume all images are distorted, you simply apply your math to them to calculate the amount of distortion. The theory is unfalsifiable, since once again the theory assumes what it is expected to prove.

In fact, I would say the data from the bullet cluster is as clear evidence as we could hope for *against* the lensing hypothesis, especially the internal lensing hypothesis. We can see that the luminosity map C already roughly matches the potential (and matches it precisely in width), so we have no need for lensing calculations. Since we can determine a width without lensing from the luminosity map, the luminosity map must not be affected by lensing. And if the width is not affected by lensing, then our original image is not distorted. And if our original image is not distorted, then lensing has been disproven. But let us study the images even more closely, to show this more clearly.

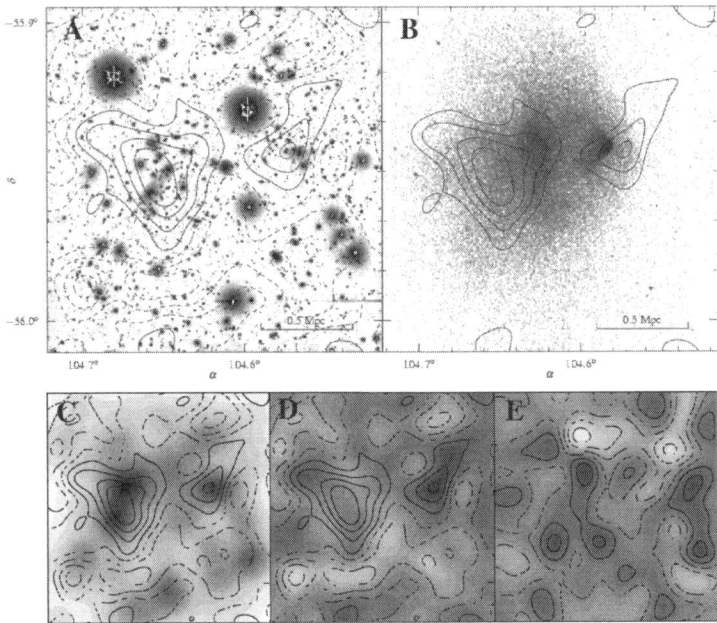

You can see for yourself that we don't need the lensing hypothesis to generate a potential map. We can generate it straight from the luminosity map. In the first paper, Clowe et al published figure C here ("the luminosity distribution of galaxies with the same *B-I* colors as the primary cluster's red sequence"), which they ditched in the second paper. They ditched it because it follows the potential map so closely, as you see. All the "stretching" away from the Chandra X-ray photo has already been done, and we don't need the lensing hypothesis. Just based on luminosity, we can see a big mismatch from the plasma field. Which means the potential is following the luminosity, not the dark matter. As width, the separation of potentials is explained by the luminosity map, and we only need to explain the change in shape. The potential centroids don't follow the *shape* of the luminous patches, but I explain that below, without dark matter.

Doesn't anyone find it curious that dark matter is supposed to be in places where we find the most luminosity? I would say it is very convenient that the dark matter map matches the luminosity map. Dark matter is dark, is it not? So dark matter concentrations would be unlikely to follow luminosity, correct? In our own galaxy, the theorists put the dark matter in a halo. But here, the dark matter is said to follow the luminosity? Wouldn't it be far more logical to propose that *shining* matter follows luminosity? And that therefore stellar matter (with its cohorts) must outweigh the plasma?

The mainstream deny this because they cannot get normal matter to provide enough mass, but they are simply missing the charge field. The charge field is the missing cohort. And of course the charge field would be expected to follow the potential of normal matter, since it is emitted by normal matter. Where you have more luminosity, you will naturally have more charge.

The mainstream might claim that dark matter follows normal matter, since the two are linked somehow. But they have failed so far to provide that mechanical or theoretical link. With charge, the link is already there. We don't have to come up with new theory or new particles or new kinds of matter. We already know that charge exists, so all we have to do is give it a non-virtual form.

But this sort of thinking is entirely too sensible for the mainstream. In paragraph 2, part 2, we find this:

Because it is not feasible to measure redshifts for all galaxies in the field, we select likely background galaxies using magnitude and color cuts.

"We select likely background galaxies." Based on what? Based on whether they fit your model or not? Even supposing that lensing were true, this entire method is slipshod in the extreme. A map of field potential like these authors are creating is a fairly precise thing, but the data we have for distant galaxies is not precise at all. Both the distances and redshifts are highly speculative, which means our knowledge of the light being bent is also highly speculative. Doubtless the likely background galaxies are chosen either because they are easier to read or because they fit the desired outcome. But neither method of selection is scientific, since it precludes obtaining an objective average. You will say that the choice is made on magnitude and color cuts, since that is what we are told, but that is only visible magnitude. We can't know which of these blurs is most or least obstructed by forward matter, so we can't know real magnitudes. This is especially true of background galaxies, which, if they were not obstructed, would not be affected by lensing. They have to be obstructed to be lensed, and then they are chosen for this lensing math because they are unobstructed. Illogical!

In part three, this is partially admitted:

The assumed mass-to-light ratio is highly uncertain (can vary between 0.5 and 3) and depends on the history of recent star formation of the galaxies in the apertures; however even in the case of an extreme deviation, the X-ray plasma is still the dominant baryonic component in all of the apertures. The quoted errors are only the errors on measuring the luminosity and do not include the uncertainty in the assumed mass-to-light ratio. Because we did not apply a color selection to the galaxies, these measurements are an upper limit on the stellar mass as they include contributions from galaxies not affiliated with the cluster.

The authors admit that the quoted errors do not include the high uncertainty in mass-to-light ratios, and they seem to think that admitting it in the body of the text is enough to make you look away. "Oh, they seem honest, so why double-check them?" But even here, they do not admit all the uncertainty. They don't even admit a fraction of it. To accept their figures, you must assume that they know how to measure such things as distance and redshift and energy and temperature almost perfectly. This despite the fact that astronomers were forced to admit just a few years ago that they were off by at least 15% in *all* simple distance measurements. As I showed in my paper on stellar twinkling, such errors are catastrophic when you come to objects like the bullet cluster. An error in distance of 15% can snowball in equations, to the point that you end up with something like an 180,000% error in final numbers. This is because all your other parameters and measurements and assumptions are also affected, so the error enters your theory in as many as 20 different ways. There I was estimating an error in the age of the universe, and this bullet cluster problem is slightly less complex than that, but small errors can still overwhelm you very fast. Even if we assume that the error in distance has been made better instead of worse in the last couple of years, we still cannot assume it has been made zero. If we were that wrong up to 2008, what makes you think we are completely right in 2010?

That was regarding distance, but the same is true of other parameters. Let us take temperature. Recently top astronomers [Tucker et al, 1998] have claimed a temperature of 17keV for 1E 0657-56. Although this temperature was later shown to be way too high [Yaqoob 1999‡], and the original method of temperature calculation was proved to be badly garbled, the higher figure has continued to be quoted. It is quoted by one of the authors of this bullet-cluster paper, Markevitch. Basically, authors quote any number they like, and get to pick and choose numbers that fit their theories. If a paper says something they don't like, they ignore it or mis-cite it.

As with temperature, so with mass-to-light ratios. The authors tell us that the mass-to-light ratio is highly uncertain, and can vary between .5 and 3, but those numbers are just estimates. If the ratio is uncertain, then the maximum variance is also uncertain, and these numbers of .5 and 3 are not firm at all. Since we are still arguing about what galaxies are made of, and have no firm data on non-baryonic matter (despite this claim of "direct detection"), it is absurd to claim that we know about the "history of recent star formation." The stack of assumptions is beginning to reach to the Moon, and therefore the claim that "X-ray plasma is still the dominant baryonic component in all the apertures" must be taken with a grain of salt. This quoted paragraph all but admits that the numbers are squishy, and since the entire paper depends on the plasma being dominant, the paper must be hanging by the slenderest of threads.

All these manipulations the authors are making are both speculative and squishy. Why not just use the luminosity map? Why ditch it and try to manufacture a dark matter map from lensing? The luminosity map is direct data that doesn't have to be massaged. It requires many fewer assumptions and manipulations to work from the luminosity map. Why hide it in the second paper?

We see this again in the implied claim that we have enough

resolution to solve this problem as the authors solve it. But since we have a sigma of only 3.4 here (first paper), we have a signal that is only 3 times the standard deviation of the noise. When I asked an X-ray astronomer for the raw data on this problem, he said this: "Unfortunately the raw X-ray data for clusters is not meaningful in the sense that the counts per pixel are zero or small, so nobody publishes the raw data." This means that the data is amenable to almost infinite amounts of pushing. The authors push again when they claim to have increased sigma to 8 in the second paper. So much improvement in so little time! This improvement is not explained, beyond our being informed that the authors used three more optical sets. But if even if sigma really went to 8, that is a measure of signal to noise, not of resolution. A blurry photo with almost no resolution will remain blurry no matter how long you expose it. These astrophysicists always talk sigma but never admit that the resolution is near zero. With only a few photons, a stronger signal is nearly insignificant. You could expose the image for a year and it would still have no resolution. To improve clarity, these images are smoothed with a 2 arcsec Gaussian, but once you do that you are no longer working with real data. It is like taking a photo with one dot per pixel into photoshop and trying to improve it by hitting sharpen. No matter what method you use to sharpen, the outcome will be a falsified blur.

None of this would matter so much if we were working straight from the luminosity map, since in that case we wouldn't have nearly as many depth-of-field assumptions. But when we start calculating lensing, we are looking at even more distant fuzzballs with even less resolution. Calculating bends in that situation is akin to witchcraft, and we can only assume that the authors must have followed the luminosity map as a guide. This may be another reason they jettisoned it in the second paper.

In the same breath that they are telling us of new optical sets, they claim that changing the cosmology will not affect the relative masses of the galaxies, but that is also false. It is only true if all masses are at the same distance, but is not true in 3D. A different cosmology will or may change depth of field in various ways, which will change a hatful of assumptions, especially in the lensing part of the theory. They just assume that the different cosmology is not very different.

In part 4, Discussion, the authors admit that another big problem with their method is that the lensing map is two-dimensional, and this "raises the possibility that structures seen in the map are caused by physically unrelated masses along the line-of-sight." To answer this, the authors give us only a probability from blank surveys, indicating that this is highly unlikely. But the bullet cluster itself is highly unlikely, if we look a blank surveys, so this means nothing. The authors also dismiss the filament proposal with more dishonest talk of probability. This is improper use of probability math. *None* of the various scenarios are likely, and we don't choose between them based on probabilities. We did not roll the dice and decide to look in the direction of the bullet cluster. We chose it based on its specialness, so the fact that it is special cannot be dismissed as a low probability.

The entire Discussion section is wildly dishonest, as it dismisses other possible interpretations with soundbite analysis and airy citations. For instance, the authors cite Markevitch et al, 2002,† as proof of X-ray plasma temperatures, despite the fact that Markevitch purposely mis-cites papers there. In that paper, Markevitch cites Yaqoob 1999 in support of his claim that temperatures reach 17keV, but if you take that link, you find Yaqoob showing temperatures of around 12keV. Since Markevitch is an author of both bullet cluster papers, we must assume that the authors don't know how to calculate temperatures or make correct citations.

All of this is worth mentioning as an example of the scruples of these authors. The form of their argument should be a big clue as to its content, which is why I have taken the time to analyze both the form and the content. Both show clearly that we are being misdirected. The photographs we have aren't that difficult to read, and are amazingly easy to explain using normal matter, which may be why these authors have to go to such agonizing extremes to keep you from seeing that. The luminosity spread, combined with the Chandra image, are all we need. We already see the decoupling, and only have to explain why the potential would follow the luminosity instead of the plasma. It is simply because the old assumptions were correct: the majority of the mass of the galaxy is stellar mass and its cohorts. The plasma is a secondary player in the unified field, as in the solar system. The plasma is *created* by normal mass and its charge, so it cannot outweigh it.

Yes, the charge field is the answer to all these manufactured conundrums. The charge field is the cohort of the stellar mass that explains everything. As I have shown in many papers, charge has mass. By the *current* equations, the statcoulomb is equivalent to mass and the Coulomb is equivalent to mass/second. This means that photons must have mass. They are not point particles and they are not massless. The have no rest mass only because they are never at rest. But they do have energy, which is moving mass, which must be included in totals.

Those who follow my own papers will say that I have presented simple equations that show that the charge field outweighs the normal matter field, so how can I say that the X-ray plasma may not outweigh normal matter in galaxies? Isn't that a contradiction? No. First of all, my charge field, which does generate or energize the plasma, is not equivalent to it. That is the first thing you must understand. A plasma is a plasma of ions. It is these ions the mainstream are weighing when they weigh a plasma, so they are still failing to weigh my charge field and the photons that compose it. The charge field is photons; the plasma is ions. Second, my charge field adds to the weight of the galaxy, but the gravitational potential won't necessarily follow it exactly. This is because the charge field is uncontained, while the matter field is semi-contained. Because of the size of the constituent particles, matter tends to orbit or condense, and this can be called semi-containment. Photons "orbit" only to a tiny degree, with enough bend only to keep them mainly in the galaxy. But photons do not condense or maintain true orbits, which is precisely why they do not constitute matter. They are too small to orbit or condense. This is why we see distant galaxies: some of the photons escape. Because they create such weak "curves", they add very little to existing potentials. Photons do not mainly work gravitationally, they work electromagnetically. They are the E/M part of the unified field, not the gravitational part. So, although they do have mass, they do not act like other matter.

This is why the luminosity map follows the potential map roughly but not precisely. We don't need lensing to generate a potential map, we need to correct the luminosity map with a charge map. Since charge acts differently than either matter or plasma, we need a third map. If we could map charge, we would find it following baryonic matter, but acting independently in some ways. Because it moves in larger curves, it runs both ahead and behind normal matter, in larger vortexes. And because it is emitted by the spin of matter, it tends to move at right angles to it, in the largest structures. This is precisely what we are seeing in the larger cluster on the left side of figure C. We see the luminosity, and therefore the stellar matter, in a barbell shape running roughly to 1 and 7 o'clock. [Notice that the highest centroid of potential is at the center of the barbell, indicating a match of center-of-luminosity and center-of-gravity.] But the total mass of that cluster runs more east to west, with a strong arm reaching out to 10 o'clock, perpendicular to the barbell. We may assume that the arm that would be reaching out to 4 o'clock is suppressed by the influence of the other cluster. We are seeing the charge field here, working at right angles to the matter field. The same thing goes for the right cluster, where the matter field is oblong mainly side to side, while the total mass is oblong up and down.

This is another reason the dark matter hypothesis is so absurd: they are seeking weakly interacting particles when they already have them. Photons *are* weakly interacting gravity particles, since their interaction is E/M not gravitational. They are so small and fast they dodge most of the gravity "field". Gravity curves them very very little.

Since curvature is a measure of potential, photons have very little potential. The mainstream is looking for massive weakly interacting particles (WIMP's), when what they should be looking for is huge numbers of very small weakly interacting particles. The added mass can be gotten either way, of course. And we already have huge numbers of weakly interacting particles: the charge photons.

You will say, "If that is true, and photons add little to the potential, then how can you propose them as mass to fill the potential gap? The original problem of galactic rotation was that we don't have enough mass to cause the velocities we see. If photons don't add potential, how can they affect velocities?" They affect velocities via the E/M field, not the gravity field. But since the E/M field is already part of the unified field, and always has been, they will affect the velocity that way. The velocity equations we have always had are UFT equations, as I have shown. Newton's equations were always UFT equations, and they always included E/M. The great thing about the charge field is that it allows us to explain planetary torques and perturbations *mechanically*. The old gravity field, either Newton's or Einstein's, couldn't explain forces at the tangent, but charge can. This applies to all bodies in orbit, not just planets. This is how photons affect orbital velocities.

You will say, "If that is true, and the equations we have always used are unified field equations, then they should have already included photons. If they already included them, you cannot add them in now!" Yes, I can, since the historical equations were incomplete. Not wrong, just incomplete. My unified field equations—for force and velocity and so on—include a second term, which expressly includes the material presence of the charge field. Newton's equations included the *forces* from the charge field, but did not include the material presence of the charge field. To say it another way, Newton included the density of the charge field as its ability to move the matter field, but he did not include the density of the charge field as its ability to take up space and cause drag. Since this term goes to zero in most cases, it can be ignored. Only when we look at very large spaces or very large forces does the second term become significant. This is what is happening in the galactic rotation problem and other similar large-scale problems. This is the unified field velocity equation:

$$v = \sqrt{[(GM_0/r) - (Gm_r/r)]}$$

Where the first mass is the mass inside r and the second mass is the mass at r. Since the second term can also stand for the drag caused by the photon field, this equation also solves the galactic problems directly.

Finally, you will say, "If charge photons replace dark matter, shouldn't we see them? Why is charge dark?" Charge is mostly dark because it is mostly outside the visible spectrum. It is the same reason we don't see most photons in the E/M spectrum: they are above or below the visible. The better question is, "Why don't we detect it?" But we do, of course. Every detection we have of everything is a detection of photons, since only the photons can travel to us. So the luminosity map itself is a detection of the charge field. We just haven't yet read the E/M spectrum data in the right way. Up to now, we have only used brightness to estimate the size of the stellar field. We have not seen that we can use brightness to estimate the local photon field as well. We have gotten used to ignoring the photon field here, so we ignore it in the rest of the universe, too. We don't realize that the brightness we see is only a tiny sample of the photons that are there, so we never even try to extrapolate up and find a figure for the photon field. Because we think the photon is massless, it never even occurs to us. But I have already shown a simple way to do it, and there are others. I have shown an extremely short method using only e, and found a charge mass inside the Bohr radius of 19 baryons per baryon. Meaning, the charge field is 19 times the matter field as a function of mass, that near to matter. This is important, since it matches the current estimate that 95% of the field is non-baryonic. This means the mainstream data is correct, and only the mainstream interpretations of that data are wrong. "Non-baryonic" turns out to be equivalent to "photonic".

Once we recognize that the charge field is the answer to all these questions, we see that the dark matter hypothesis rests on a contradiction. To solve the problems it claims to solve, this dark matter must be non-baryonic, which means it isn't like normal matter. How is it said to be different? It is mass that is weakly interacting, which means it must be mass without density. But that is to ignore the definition of both mass and density. Mass is *defined* as density times volume, so you cannot propose mass that is weakly interacting in this way. Do these people mean to jettison M=DV from the textbooks? They want you to think they are remaining true to Newton (unlike their MOND enemies), but they throw out a fundamental equation from chapter one of basic physics. They claim they are theorizing "independent of assumptions regarding the nature of the gravitational force law", and then propose mass without density. They want you to think they are proposing a new kind of matter, but they are really proposing a new kind of mass, one that is redefined as *not* a function of density. Just as they want the void to be nothing one instant and something the next (symmetry breaking, Dirac's field, Higgs field), and just as they want a particle to have energy one instant and none the next (virtual particles), they want mass that has extension and no extension, energy and no energy. They want non-baryons that have mass when we are creating potentials in distant galaxies, but that have no mass when they are traveling through the Earth. If these non-baryons have potential, then they MUST interact with other particles that also have potential. Otherwise the words have no meaning. The same particle cannot act as a gravitational entity in a distant galaxy and as a non-gravitational entity when flying through the Earth. These theories are flagrantly illogical.

When Bishop Berkeley was criticizing Newton's calculus proof, he called Newton's fluxions "the ghosts of departed quantities," since they were sometimes zeroes and sometimes infinitesimals. These non-baryons of the dark matter hypothesis are the same sort of souls of departed protons, circling in haloes about galaxies like a far-flung hell of moaning spirits. Where else should these ghosts of dead particles go but into the outer regions, far from the sight and influence of good people? We may assume shattered souls of particles and anti-particles are transported here after collision, perhaps by a battery of non-baryonic angels tooting on non-baryonic trumpets. As with the hosts of Heaven and Hades, these ghosts will be weighed after the judgment, but not on the scales of the Earth.

As you now see, the photon solves all these problems, as

long as we give it mass and radius. The photon, though it has mass, is not massive, so it is naturally low density. Even with a mass of 19 baryons inside the Bohr radius, it is still very low density. What density, you may ask? We can now calculate it. Since volume depends on radius, the volume inside the Bohr radius is 237,000x the proton radius. But we have 19x more mass, so the density of the charge field is 12,474x less than the matter field. And there you have it: high mass, low density, just like the WIMP people are looking for.

We may conclude by noticing that these authors and all authors of mainstream papers never do what I have done here. They never bother to say precisely what their theory is. They never get down to brass tacks and try to uncover all the mechanics involved. Their so-called discussions are more deflections or misdirections than discussions, since they mention a few of their colleagues or reviewers caveats but never bother to mention their opponents' hardest questions. They never lay it all out on the table and let you decide for yourself. They hide and massage data, and often manufacture it. They are not above dressing not-A up as A, and trying to walk it by you. You should learn to catch them at this, as I have. You should learn to see through the veils. Neither you, nor physics as a whole, will make any progress until that happens.

Postscript: Since around 2008, the mainstream people involved in the dark matter hypothesis have dropped the bullet cluster as a lead story. It may be because they realize it is a sitting duck. Joe Silk, the Savilian chair of astronomy at Oxford, admitted at a recent colloquium (Oct. 2010) that the current bullet cluster theory was "a mess". Rather than let that deter him, however, he continued to pursue his thesis, which was that supersymmetric particles may be a good candidate for dark matter. In other words, he is making up particles to fill holes that he and his colleagues made up. Problem is, supersymmetry is a theory with no data, created to explain another theory with no data, which itself was created to explain a non-mechanical theory made up from whole cloth, again with no data and no possible data. Silk even admitted that supersymmetric dark matter has 123 adjustable parameters. He stated this with some pride, as current physicists do when they are trying to wow you with the math. A math with 123 adjustable parameters is not an albatross, it is a prize goose. That is, it is a prize goose as long as it is your goose. When the opposition has a math with adjustable parameters, THAT is an albatross. Silk proved this double standard when he dismissed non-dark matter gravity theories (such a MOND) as having "too many adjustable parameters." So much for consistency as a fundament of science. He also said MOND makes no predictions. But of course dark matter theorists have never made a prediction. Dark matter is not a predictive theory, it is a stop-gap theory, as are all current theories.

Silk also admitted that we know almost nothing about dark matter, except that it is affected by gravity. But in answer to why, if galaxies are predominantly composed of dark matter, the central black hole is not dark matter, Silk replied that dark matter can't form black holes because it can't lose enough energy. Apparently dark matter is gravitational, except when we want to exempt it from acting gravitationally.

*http://iopscience.iop.org/0004-637X/604/2/596/pdf/59234.web.pdf
**http://arxiv.org/PS_cache/astro-ph/pdf/0608/0608407v1.pdf

†http://iopscience.iop.org/1538-4357/567/1/L27/fulltext
‡http://iopscience.iop.org/1538-4357/511/2/L75/pdf/985488.web.pdf

Chapter 9

The
COSMOLOGICAL CONSTANT
<u>is</u> the CHARGE FIELD

Part 1

Everything I will say here is contained in my other papers, but I find that it is best to be explicit. I take nothing for granted anymore in these papers, and constantly remind myself that it is necessary to make clear every new link that my theories contain. I have already proved that dark matter is in fact the charge field [chapter 7], and the mainstream now proposes that the cosmological constant is caused—at least in part—by dark energy or matter. So a good reader will already understand that I am proposing the charge field as replacement for both dark matter/energy and the cosmological constant. But in this short paper I will explain exactly how the charge field got lost in the evermore complex field equations of the 20th century.

The problem began with Newton, who supplied us with the first modern field equation. His gravitational equation $F=GMm/R^2$ is both the starting point of the problem and of my solution to it. As beautiful as Newton's equation was admitted to be, it was known to fail very early on. It had to be fluffed and pushed by Euler/Lagrange/Laplace in the 18th century to match solar system problems at the time. This pushing, because it was relatively small, was considered to be a confirmation of Newton's gravity theory, and it was sold as such, but the top physicists and mathematicians of the time were not so sure. Because they weren't able to correct the field at the foundations, as a matter of mechanics, their uncertainty kept to the level of unease, rather than rebellion. They saw their job as supplying the mathematical extensions necessary to keep the field alive, and they did their job pretty well.

For over a hundred years celestial mechanics existed like this, with some other minor updates in the 19th century. But it wasn't until the arrival of Einstein in the early 20th century that the next big step was made. To solve a pair of fairly subtle problems (Mercury's perihelion precession and Solar deflection of starlight), Einstein added his relativity transforms to the existing field equations. But that turned out to be just a lead-up to a pair of far greater problems. Einstein applied his field equations not just to limited expanses of space, as had been done all along, but to the universe as a whole. Using the naïve gravity-only interpretation of the field that had come down to him directly from Newton, Einstein thought this meant his universe should be shrinking. You see Einstein had extended Newton's equation, but he hadn't really corrected it. Beneath the relativity additions, Newton's field was the same as it ever was: gravity-only. With gravity only, the field taken as a whole should shrink. In fact, it should shrink exponentially over time, and shouldn't even be here now. The universe should have shrunk to nothing long ago.

To correct this, Einstein added a constant to his field. In the beginning Einstein didn't have any data to match, so he just chose his constant based on a hunch. He figured the universe was stable, and chose a constant that would perfectly offset gravity. In the 20th century, this choice was seen as Einstein's greatest blunder. He called it that himself. He said that he should have waited for some data—and that wouldn't have taken long since Hubble was right around the corner, so to speak. But what I have discovered is that this interpretation of the events misses the point. This whole "blunder" doesn't really matter, and the crux of the problem has been missed up to now. You see the problem is that everyone from Newton on down has just assumed that the field was gravity-only. The field equations express a total force, that force is a single force, therefore the field must be a single field, right? Wrong. A single force does not in any way imply a single field. A compound field, made up of two completely separate fields, would also create a single total force. And this is in fact the case. I have shown that Newton's equation contained two fields from the beginning, the charge field being the hidden field (hiding within G). This is why Coulomb's equation looks just like Newton's in form: they are both unified fields, the charge field being hidden in Newton's equation and the solo gravity field being hidden in Coulomb's equation (hidden within k).

The reason this is important here is that Einstein, not knowing that, interpreted his own new field equations as gravity-only. Like everyone else, he couldn't pull apart Newton's field equations. He didn't even try to pull them apart, since he had no idea Newton's equations were unified or compound. Yes, Newton's field equations *already* contained the charge field, and the charge field is *already* in vector opposition to the solo gravity field. Because Einstein's field equations contain Newton's field equations, Einstein's field equations also contain charge. And that being so, the "cosmological constant" was already inside the equations. Einstein didn't need to add *any* constant to the field equations, he just needed to understand the mechanics that the equations already expressed.

What this means is that *all* values for the cosmological constant have been wrong. No single constant can fix or extend the field equations in the right way, since the constant is misdefined from the beginning. Since it is the charge field that opposes solo gravity, and since the cosmological constant is meant to stand for this opposition, the only way it could work is if the constant stood for charge. But if you let the constant represent charge, you can't integrate it into the existing field equations. Why not? Because the existing field equations already contain charge. It is mathematically impossible to fix an equation by integrating (a second time) a variable the equation already contains. Logically, the only way to fix the equation is to fix the variable that already exists. In other words, if we represent charge by the letter "c", and we find that our initial field equation containing c is not working, we don't leave the original equation alone and try to add or multiply by c a second time. No, we rewrite the original equation so that c has the proper relationship to the other variables. That is what I have done in my unified field equations. I have shown where charge exists in the old equations and how to expand the equations so that they answer more data.

Soon after Einstein's "blunder", Hubble made his big discovery and a second huge problem loomed. Einstein, having learned from his mistake, and being somewhat more scrupulous than the rest, refused to simply boost his constant up to match what was thought to be the new expansion. He saw that there was something wrong with the field equations at a fundamental level, and he set to work to find out what it was. He saw, first of all, that the field equations wouldn't unify with the quantum field. That was a stunning discovery for him, because he knew that he had patterned his field on the E/M field of Maxwell. He had even used the motion of light as his first postulate. The two fields should have fit together like a child's puzzle, and yet they were not only difficult to reconcile, they were impossible. How could that be?

His contemporaries were not so hard on themselves, or so rigorous, and they had no problem changing the constant to match new data. They have been doing it ever since. The 20th century physicist, even the theorist, was a specialist, and specialists don't bother so much with the big picture. Unification has been a sexy topic for the magazines, but it hasn't bothered too many physicists too much. Particle physics has dominated theory, and particle physicists don't care a fig for unification. Celestial bodies such as Mars are just probabilities like electrons, and don't exist until we look at them and force them to decohere, so why worry about unification (see Murray Gell-Mann's *The Quark and the Jaguar*). Macro-objects are just the ghosts of wave mechanics, and you don't worry about unifying with ghosts.

But the answer Einstein was looking for was that both his own field equations and the quantum field equations were *already* unified. So you didn't need to stack them, and couldn't stack them. It was never a matter of stacking the fields, it was a matter of locating gravity in QM, and charge in the field equations of GR. This is what I have done. I have shown where gravity exists in the quantum equations and where charge exists in Einstein's equations. This effectively unifies the two fields.

Part 2

I could end there, but I will now extend these comments in a second part. There are a few things left to be said that may answer questions some still have. Currently, celestial mechanics is split by two contradicting interpretations. When we are taught about the solar system, we are taught stability. We aren't even taught that the asteroid belt was caused by a collision anymore: that is too much instability too close at hand to stomach. Velikovsky scared everyone, and they are now huddled under the illogical assumption that the solar system was always pretty much what it is now. But when we are taught about the universe, we are taught instability. The universe is expanding at a fantastic rate, we are told, and will end up as nothing more than a mist. Since the time periods are longer there, we can learn that without too much anxiety, apparently. Unfortunately, *both* interpretations don't fit the existing field equations. Those equations are still gravity-only, and a gravity-only universe can only shrink. In that sense, Einstein is still right: if you are going to stick with gravity-only, you can explain stability or expansion only with a fantastic mathematical fudge like the cosmological constant. If you only have one force field, and that field is a pull, every addition to your field, be it mass, energy, or a constant, must be a fudge. For instance, if we accept the current claim that the cosmological constant is dark energy or matter, we still have a contradiction. By definition, any matter or energy must have mass or mass equivalence, which means it must enter the current field equations as a gravitational entity. In other words, we are told that 95% of the mass of the universe is dark, but that means that it is not only dark, it is *mass*. If it is mass, it must be gravitational. Well, if it gravitational, it cannot also be anti-gravitational, can it? The cosmological constant is supposed to balance or over-balance gravity, right? How can 95% of the mass of the universe be anti-gravitational?

You see, to propose that any part of dark matter/energy is the cosmological constant, new physicists must be proposing that dark matter/energy is anti-gravitational, which is simply a contradiction. By both the definitions of Newton and Einstein, mass, matter, and energy are those things that cause gravity, so they can't offset gravity by any amount. Mass and energy must go the gravity column or the gravity side of the equation.

These physicists try to weasel around this by claiming that dark matter/energy is "weakly interacting", but to act as the cosmological constant, the dark matter/energy would have to be *negatively* interacting gravitationally. The cosmological constant isn't "weakly interacting", it is a repulsion. They have never explained how that can be.

They will say, "OK, but your photons also have mass. So we return the question, smartass! How can photons be anti-gravitational?" I have never called photons anti-gravitational, but photons make up the charge field, and the charge field is in vector opposition to the solo-gravity field, so I suppose you could call them anti-gravitational if you like. They certainly do create a repulsion, since the photons in the charge field are emitted by particles and bodies. They are moving out from the surface, and therefore oppose the gravity field. It works like this: we start with solo-gravity, which still creates a vector in, in my field. I have changed nothing there. Then we add the charge field, which is just a photon gas, in the first instance. I have changed nothing there either, since we already know about the spectrum. I just take the existing and known spectrum and use it as the charge field. Then I give all atomic and subatomic particles spin. I change very little there, since they already have several spin quantum numbers. I just make the spin real. Then I propose that these particles recycle the charge field, by actually taking the photons in and re-emitting them. In this recycling, they simply obey angular momentum rules. The particles have more angular momentum at the equator than at the poles, so photons go in at the poles and out at the equator.

I will be told, "To make this recycling stable over time, the same number of photons have to go IN as OUT, so the IN vector must cancel the OUT vector. Your charge field can't sum to OUT, which means it can't be in vector opposition to the gravity vector." But that is false. The same *number* of photons go in as out, but the energy in doesn't equal the energy out, again due to angular momentum. Because there is more angular momentum at the equator, the out-photons have more energy than the in-photons. This creates and maintains a total field vector OUT, which balances the gravity vector.

I will be answered, "But what maintains your spin? Such a situation in the macro-world would cause the spin to stop.

You have said your photons have real mass and size, so they cannot be emitted without a loss of energy of the emitting particle." False again. It is the angular momentum differences in the sphere that cause the energy differences, and those same momentum differences cause the spin to maintain, given a sea of photons. Density differences in the external photon field, just beyond the sphere, are naturally created by the recycling process, and these density differences act as potential differences, maintaining the spin. In other words, the energy in the photon field is transmitted to the spinning particle in such a way as to maintain spin, and it is precisely the spherical shape that allows it to do this.

I will be answered, "In that case, the photon field must lose the energy it gives to the particle, to maintain that spin. Over time, the photon field would dissipate." False again. The photon field loses energy as it goes in, yes, but gains energy as it goes out. So we have a cycle. The particle borrows energy from the field, with which it maintains spin; it then gives that energy back to the field with the higher angular momentum at the equator.

I will be answered, "You have just contradicted yourself. You said that there was more energy out, to explain the vector out. Now you say that the energy in and out are the same, to conserve energy." No, I haven't said anything about conserving energy. The charge field, by itself, *doesn't* conserve energy. Only the unified field conserves energy, and I haven't included gravity in my tally yet. As I have shown, the photon field doesn't lose energy by spinning the particle. In fact the reverse. The photon field *gains* energy from being recycled. So the particle doesn't burn photons like a car burns gas. The recycling actually gives the gas more energy.

You will ask how this can be. We have a double energy production here, with the particle getting spin and the field also gaining energy. Where does all that energy come from? From gravity. Gravity acts as a well of free energy in current theory, and I also don't change that. In my theory gravity remains a well of free energy, and when I say that the unified field is conserved, I only mean that the input from the well remains constant.

Current theory also assumes that gravity is a well of free energy, despite its claims to the contrary. Here on Earth, the "pull" remains constant, and that is the only way that energy can be said to be conserved. Neither universal nor local energy ever sums to zero: that was never the meaning of conservation of energy. Conservation of energy just means that gravity neither gets greater nor lesser, it doesn't mean that gravity sums to zero with some other force.

I will be told. "That is exactly what conservation of energy means: a sum to zero of a system. Yes, the Earth's gravity is a constant input into the universe, but it is balanced by other bodies and their gravity, so the system does sum to zero." Maybe, maybe not, although if current theory is correct, and we dissipate into a mist, energy will not have been conserved, will it? It will have been lost. But what I am talking about is something different. There may be a sum to zero if you mean that all energy that enters the universe via gravity will be used up. It may be that the free well of energy that is gravity is used in full, always, which is what causes a conservation of energy. But even so, energy isn't conserved in the sense that the free well of gravitational energy is not *restored* by any circular process that we know of. Where does the Earth get the energy that it transmits as gravitational? No one knows. No one has shown a *process* by which that well is filled. That is what I mean by conserved. By this way of looking at it, the total energy of the universe is not conserved and does not sum to zero. At any one moment, the energy input into the universe is hugely positive, and over time, it is even more positive, since we have to add up all those moments. There is no sum to zero.

So you have seen that there are several ways of looking at conservation of energy, at stability, and so on, and they shouldn't be confused. In my mechanics, we have no universal conservation of energy, though we may have a universal conservation of energy *levels*. And my vector opposition of gravity and charge, although it is responsible for the stability of orbits and other things, implies nothing about the universal conservation of energy or any summing to zero. As I just showed above, it is mainly a way for gravity to seem to resist itself, by which one energy input into the universe can split into two opposing fields.

Why do I say that we have only one energy input, but two fields? Because the charge field is not a constant input of energy like gravity is. The charge field is just a photon gas, with a constant energy level. But gravity is not a constant energy level, it is a constant energy input. It is like the difference between a velocity and an acceleration. The charge field is the velocity, and the gravity field is the acceleration. Each photon has a velocity, and a velocity requires only an initial input. It does not require a constant input. But gravity requires a constant input of energy from somewhere.

You will say that if photons have real mass, and if the field has real density, its energy level cannot be maintained without an energy input. True, but in my mechanics, this input is relatively small compared to gravity, and it in fact comes from gravity. The energy of the photon field is maintained by the mechanism above, and gravity drives that mechanism. In other words, we had the photon field gaining energy by being recycled, above. And, I have just been reminded that the photon field will also lose energy due to being real (due to collision with non-photons). I suggest these gains and losses may balance, and this balance is what keeps c stable.

In the same way, particles larger than photons are kept spinning by this same cycle. The recycling of charge, along with solo gravity, together create a feedback mechanism. The two fields, being in vector opposition in the vicinity of all particles, create a loop or a tension which is capable of driving the various quantum processes. By this I don't mean anything esoteric like loop quantum gravity. I mean a loop in the sense of a simple cycle. The recycling of the photons is what creates the real circle of energy, and we can even draw the circle on a simple diagram.

At the most basic level, this circle is created by the spherical shape of the spinning particle. It is the sphere that naturally creates the field potentials, via the poles and the equator and the large angular momentum variance. This is why I have resisted the fancier particle models that seem to be in vogue. And I don't just mean the exotic shapes proposed by string theory. Almost all alternatives to string theory incorporate exotic shapes with fancy names. In my opinion, this is just a public relations move. Fancy new shapes and terms help generate interest for new theories. But we have ample proof of nature's preference for the sphere. Just look around. The planets and stars exist in the charge field, just like electrons do. If the sphere is good enough for stars, it is likely good enough for electrons. Regardless, it is best to use the sphere as the default shape in the early part of major theory correction, in my opinion, since it vastly simplifies the math. Once we get the fields sorted out, we may be in a position to take a closer look at shapes. Until then, arguing about shapes is like arguing about paint samples or throw pillows before you have the walls up.

Chapter 10

How to Calculate the
EARTH'S ECCENTRICITY
Using the Charge Field

In a series of other papers, I have calculated the axial tilt of many planets, the eccentricity of the Moon, the Bode series, the magnetopause of both Earth and Venus, and many other numbers using my new unified field equations. Here, I will calculate the Earth's eccentricity. The Earth's eccentricity is said to be caused by gravitational perturbations from other bodies, but once again both the math and the theory fail. It is interesting to go to Wikipedia on this one, which normally has all kinds of math on things like this. This time we get nothing. Wiki doesn't even give us a section on the cause of orbital eccentricity. It doesn't even mention that eccentricity must have a cause. Eccentricity is used as a cause of climate, but it doesn't have a cause itself, I guess.

The Earth has a low eccentricity of .0167, about 1/3 that of the Moon. If we try to explain that with gravity alone, then we must look mainly to Jupiter and Venus. All other perturbations will be small relative to those. According to current math and theory, Jupiter should be the main influence, with a maximum force almost twice that of Venus and 26 times that of Mars. These perturbations are at closest approach:

$$\Delta a_{Ven} = GM/R^2 = 2 \times 10^{-7} \text{ m/s}^2$$

$$\Delta a_{Jup} = GM/R^2 = 3.6 \times 10^{-7} \text{ m/s}^2$$

$$\Delta a_{Mars} = GM/R^2 = 1.37 \times 10^{-8} \text{ m/s}^2$$

$$\Delta a_{Sun} = GM/R^2 = 6 \times 10^{-3} \text{ m/s}^2$$

But we can't even add the effects from Venus and Jupiter, since when the two are in line at closest approach, they *subtract*. This means the maximum perturbation is by Jupiter and Mars in line, at 3.7×10^{-7} m/s^2, and the minimum perturbation is 0, when Jupiter, Venus, and Mars are stacked behind the Sun. The difference between maximum and minimum should give us the eccentricity, and that is still 16,000 times less than the effect from the Sun. There is no way that can cause an eccentricity of .0167.

Put simply, the gravitational math is all magic. It is fudged from top to bottom. I have already shown you in other papers how the Moon's orbit is a mess of fudged and unsupported equations. If the Moon's orbit is fudged, why would anyone trust the math of perturbations between Jupiter and the Earth, or Venus and the Earth? If they can't solve the Moon's orbit, then they can't solve anything.

The Moon is orbiting the Earth, so there is an apparent gravity field between them. But there is no gravity field between the Moon and Venus. Venus is not perturbing the Moon gravitationally, or the reverse. They only influence one another via charge. This is what Newton, Kepler, Laplace, Lagrange, and Einstein could not comprehend.

They had equations that contained this information, as I have shown, but they could not unlock them.

Rather than look at the math again, let's look at the field again. The top physicists brag on their T-shirts that they have gotten rid of force at a distance, although you wouldn't know it from perturbation theory. If there is no force at a distance, how is Jupiter perturbing the Earth? Don't tell me gravitons, since 1) they haven't been discovered, 2) they aren't logical to start with. You can't explain pulls with fields of particles. Beyond that, many of these perturbations we are given by the mainstream are pertubations at right angles or tangents. The historical gravity field is centripetal: it can't provide such forces or accelerations or perturbations, and both Newton and Einstein agreed on that. They said it out loud, in simple declarative sentences. I will be told that space is curved, and that perturbations are geometric, not dynamic. But that explains nothing, especially not orthogonal perturbations. Gravity can't curve enough at such distances to provide these perturbations. Einstein's space is curved around a body. As such, it can explain orbits (sort of), but it still can't explain straight line forces between bodies that aren't orbiting one another. A perturbation between Jupiter and the Earth is a straight line force, since the Earth isn't orbiting Jupiter, or the reverse. The Earth isn't in Jupiter's curve, or the reverse. The Earth is only in the Sun's curve.

Remember the rubber mat and the ball bearing they like to show you, to help you visualize curved space? Well, that explains the curve of the orbiter only, and thereby the appearance of a force. What curved mat is between here and Jupiter? And if Jupiter perturbs the Earth and the Earth also perturbs Jupiter, which way is the mat curving? Can it curve both ways at once? In other words, can space be concave and convex at the same time? No, it must be one or the other.

Einstein didn't really explain anything about orbits, either. He just gave physicists two theories to ignore and corrupt instead of one. Without the charge field, neither Newton's equations, Einstein's equations, Laplace's equations, nor Lagrange's equations can explain the motions we see. Yes, maybe Lagrange came closest, since he was the best cheater. He saw that we needed a differential equation, with gravity somehow resisting itself, and he cleverly pushed the math to create one. But the Lagrangian is just as smelly as all the rest of the historical field math, and it is way past time we threw it out and started over.

I will do that now. I have already shown (in my paper on Axial Tilt) that the Earth's tilt is caused mainly by charge forces from the four big outer planets (Jupiter, Saturn, Uranus, Neptune). It turns out that the same thing causes the eccentricity of the Earth's orbit. You will say that if they are caused by the same thing, they should show the same percentage changes in the field, so that the amount of eccentricity should match the amount of tilt. They don't match, therefore I can't be right. Well, this is why no one has solved this, but it can be solved quite easily. The tilt and eccentricity do match, as I will now show.

I will once again limit my math to the four big planets, simply estimating an answer. I showed that the four planets were responsible for a tilt+inclination of the Earth of 33.66*, which, being 37.4% of 90, indicates a 37.4% difference between the planets' effect and the Sun's. Since the planets are an average of 23 times further away from the Earth than the Sun, we divide .374 by 23 to obtain .0163. That would be the eccentricity just from the four planets and the Sun. Since the actual eccentricity of the Earth is .0167, I am very close already.

To understand how my simple math worked, we have to go back to the ellipse, as proposed by Kepler. Kepler proposed an ellipse with two foci, the Sun being at one focus. But that isn't the way real ellipses are made. There is no body at the other focus, for example, so the forces in a Kepler ellipse are ghost forces. They don't exist. The second focus in the Earth's ellipse should be outside the ellipse, as we recognize when we start doing the actual perturbations. As I just showed, the second "focus" is the average distance of the four big planets. It is sort of their orbiting center of mass. Since it is at 24AU, it can't be inside the Earth's ellipse. This would be true even in current theory, since whether you are calculating gravitational perturbations or charge perturbations, you can't propose that the source of your perturbations is coming from within the Earth's ellipse. Yes, Venus' perturbation is inside the ellipse, but all the other important ones aren't, and the average perturbation certainly isn't. Therefore the second center of force can't be at Kepler's second focus.

I will be told that you can solve with barycenter forces, since the center of mass of the four orbits can be found, and it will be found to be at the other focus, inside the Earth's ellipse. But even if that were true, and even if the problem could be solved with gravity, that would be a mathematical solution only. Again, no real body exists at the barycenter, so no real force or effect emanates from there. It is better to attach the math to the real bodies, so that the real influences can be seen and studied. These problems I am solving persist to this day precisely because physicists have given us mathematical instead of mechanical solutions. We don't need to know how abstract or geometric ellipses are created, we need to know how *physical* ellipses are created. Geometric ellipses are created with two interior foci, but celestial ellipses clearly are not. Celestial ellipses are caused by forces outside the circle.

Besides, the gravitational field can't solve this one, so the question is moot. I might be able to solve by finding the barycenters of charge, in a like way, but I find that pointless. It is much preferable to describe the real mechanics.

You will say, "But didn't you just find a center of mass to solve?" Kind of, but that is not a mathematical trick, that is just an average distance. You take the four radii and divide by four. It is not a center of mass, like a barycenter, it is an average orbiting distance of mass. The force is still coming from outside the Earth's orbit, so the direction of the force is clear. I have not moved the second "focus" inside the circle to misdirect you.

This means that the ellipse of the Earth is not caused by a second pull from within, it is caused by a second push from without. The big planets are pushing on the Earth with their charge fields, with their photons.

You will say, "Above you showed that the eccentricity was just 1/23 times the tilt. But you already used the number 23 to find the tilt, in that paper. How can you use the same number again? First you find the planets have their charge compressed by distance, which means you multiply by 23; then you divide by 23 to find the eccentricity. You seem to be going around in circles. I am lost!" Well, we have to look at the way eccentricity and tilt are measured. The quickest way to make you see this is to remind you that if the force outside the circle were constant, there would be no eccentricity. Remember that we are explaining the ellipse now by a perturbation from outside the circle, not a second focus inside the ellipse. If the force from outside (by the big planets upon the Earth) were constant in strength and direction, this would only change the radius of the orbit; it would not create an ellipse. To create an ellipse requires a *varying* force from outside. You can see this in the way I did the gravitational perturbations above. We calculate the variance from maximum to minimum. So the number .0167 is a measurement of this variation. In other words, it is not telling us the difference between the charge field of the Sun and the charge field of the four planets. It is telling us the *variation* in this difference. So the tilt is a measurement of the difference, and the eccentricity is a measurement of the variation in the difference. This is why they aren't the same number. One number is the *change* in the other number.

The reason we can use the number 23 to go from tilt to eccentricity is that the number 23 tells us the variation in the number from the planets. In seeking the variation, you seek the difference between the maximum and the minimum, right? Well, we have found that 23AU is the average distance of the four planets from the Earth. How do you calculate an average? You add the four distances and divide by 4. How do you find a difference between maximum and minimum? Maximum is when the four planets align and

157

minimum is zero (when the four planets are stacked behind the Sun). When the four planets align, you find the charge distance by adding the four distances and dividing by 4. *The average distance and the variation are the same number!* The average distance is 23, and the variation in charge difference is 23-0. So the first time I use the number 23, it stands for average distance. The second time I use the number 23 (as above), it stands for variation. The number is the same only because the math is very similar for average distance and variation.

Doubters will say, "That math is beautiful in a strange and curious way, but it doesn't make any sense. If charge toward the Sun increased with distance, then things even further away would have even greater charge. The Oort cloud would run the Solar System." But I have never said that all distant objects have their charge increased by this method. It only applies to objects in orbit around the Sun. It is a unified field effect, so the object in question has to be IN the Sun's unified field. Its charge must be captured by the vortex of the Sun, and this vortex has a limit. I haven't been able to calculate yet where this limit is, but it is probably somewhere just beyond Pluto. Farther than that, the charge is not channeled toward the Sun in an efficient manner. In other words, charge *out* from the Sun will have dissipated so much that it cannot channel charge *in*. Remember, we have charge moving in both directions. It is doubtful, for instance, that much charge from the Oort cloud makes it back to the Sun. But, yes, the charge that does make it back will be compressed and its density will be increased, increasing its effect.

You will now say, "But in other papers you have gotten rid

of positive and negative charge. For you, all charge is positive. How can you have charge going in both directions? How are the potentials created?" They are created by poolball mechanics, not strange charge mechanics. In other words, it is not mathematical potentials, created out of nothing with plusses and minuses, that cause the charge motion, it is simply the recycling of the field. As I have said in other papers, you have to think of the charge field as wind, not as mysterious potentials. Basically you have two winds: you have the incoming wind of charge from the galactic core to the Sun. The Sun takes in this wind at the poles, due to spin, and emits it at the equator. The emitted wind moves opposite to the incoming wind, but the two are interpenetrable. The photon wind is so fine, as a matter of particles, that it doesn't interfere with itself much. Photons do collide, but the collisions only affects the spins, not the linear velocity. Which means that only the magnetism is affected by these collisions; the linear charge (which causes electricity) isn't. Although photon fields are mostly interpenetrable with eachother, they are not interpenetrable with normal matter in planets. Since baryons are quite large compared to photons, we have many orders (almost 12; see the number G) more collisions, and these collisions create measurable forces or motions. The sum of all these collisions is what causes perturbations.

Now that we see how the ellipse is really created, we see that the big planets are not only the cause of the inner ellipses, they are the reason these ellipses are so small. The fairly large charge forces from the outside keep the Earth's orbit constrained. Even if it were bumped into a greater eccentricity by an impact, the charge field would resist this eccentricity, and would tend to push it back to a lower eccentricity over time. This is why even large impacts are rarely fatal to an orbit. Planets and moons will break up before they will crash into a primary or eject from the system altogether, as we see from the debris around Jupiter and Saturn, as well as from the asteroid belt.

In closing, I will answer some more questions. A reader has commented on my tilt math in this way, "I have studied your math from that other paper on tilt, and it appears that you have found that the four big planets have a charge effect greater than that of the Sun. How is that logical? Look here, according to you, altogether the four big planets have a charge .0014 that of the Sun, and an average distance of 23 times more than the Sun from the Earth. You have calculated in your paper on Mercury that the Sun has a charge of 796. So the fourth root of that is 5.93. That is the charge at the Earth. The charge of the big planets is 796(.0014)23= 25.6. The planets would appear to have 4.3 times as much charge as the Sun."

Well, that math isn't correct, but this reader is correct that the planets have a large charge density and a large field effect, both on tilt and eccentricity. I have already explained this in part 2 of the tilt papers. In looking at perturbations, we are looking at charge density, not charge strength. Yes, the perturbations from the planets, especially Uranus and Neptune, have very high charge densities in the math, since they are compressed by the long distance. This gives them great power as perturbers, but it does not mean they are responsible for greater overall charge than the Sun. You will say that charge density should be a measure of charge strength, and in some cases that is so. But here it isn't. A body with greater charge density would have greater charge strength than a second body only if equal amounts of charge are arriving at the spot in question over the same time. But that isn't the case with the Sun and the planets. The same amount of charge is always arriving from the Sun to the Earth, but the amount of charge from the planets varies according to positions. Therefore, over the course of a year, say, the Earth gets much more charge from the Sun. But because that charge is steady, it doesn't *perturb*. It doesn't cause a change because it is the baseline, and a perturbation is a *change* from normal. The high density charge perturbations from the planets peak about four times a year, but that is plenty to cause a change in the tilt or the eccentricity. Since celestial bodies are very large, with large inertias, you do not have to have continuous forces to create perturbations. Four peaks a year is more than enough.

But let us use my reader's math to rerun the equations, perhaps making them a bit clearer, especially for those who haven't read the tilt papers closely. He has tried to mirror my math, but hasn't quite got the feel for it yet. He has actually done a bit too much math, but we will use his numbers to get the right answer. If the planets have a charge .0014 that of the Sun, then we only need to multiply that by 23 to get .032. Since the Sun has a charge of 1 in that math, the number . 032 is already the number we need, without bringing in the real charge number of the Sun. You see, the number .032 is already a fraction of 1, so we don't need any other manipulation. We just multiply by 23 again to include the field variation, to get .736. Since I have shown in another paper that the Sun's number at the Earth is .163, we find the number for the big planets to be .122. That's a lot of charge density, admittedly, but it is not more than the Sun's, not even at maximum perturbation.

And so I have proven my point above. I said that gravity perturbations 16,000 times less than the main orbital forces could not create an eccentricity of .0167. We have just seen that it takes a variation in charge density almost 75% that of the ambient charge to create that much eccentricity. In other words, the charge from the big planets is only 25% less than the charge from the Sun, as a matter of density. Not 16,000 times less, but ¼ less. The current math isn't even close. Currently, it is thought that eccentricity is low because the perturbations are low. But, as we have just seen, the eccentricity is low because the outside influence is near the inside influence. It is the low percentage change that causes the low eccentricity, not the low relative value of the perturbation.

Let's see how this fits into the unified field at the Earth. We know that the Earth must be balancing charge from both directions, since the actual distance of the Earth from the Sun can't be explained by forces from the Sun alone. Let's return to the numbers. The gravity of the Sun is 1070, which drops off by radius. So at the Earth this number is 1070/23,456=.0456. The charge is -.163. The given velocity can't balance those numbers, since it adds an acceleration of only .0048. The Earth isn't anywhere near where it would be with the Sun alone, and no other planets. It would be a good deal farther away, to get that charge number down. This is logical, because it is the large outer planets that are pushing it back in to its current radius. According to this simple math, the big planets should be pushing with a combined charge acceleration of .163 + .0048 - .0456 = .122. As you see, I have balanced the unifed field once again, showing all the factors.

162

This brings up more questions. You will say, "If the Sun's gravity drops off by radius, you should use the Sun's radius to calculate the drop off, not the Earth's. And yet you use the number 23,456, which is the number of Earth radii in one AU. How does that make sense?" Well, the number 23 above was found using 1AU as the baseline distance. The number 23 means 23AU. It does not mean 23 solar radii. Since we are comparing forces from the planets to forces from the Sun, we must use the same distances for both. The numbers won't be comparable, unless the distances under the numbers are the same. The number .122 comes from using the number 23, which comes from using 1AU, you see. Since I am comparing .122 to .163, I have to take that into consideration. And since .163 was found using both 1AU and 1 Earth radius, I have to use 1 Earth radius in the Solar gravity number as well. You may need to consult the math in my Lagrange point paper to comprehend my point here. I am combining numbers in three different fields here (Solar gravity, Solar charge, planetary charge), and combining them at the Earth, so I have to scale them all to the same baseline field.

*The Earth's actual tilt to the Sun's equator is 30.55, but we have to include Mars and Venus to find that number.

Chapter 11

Where is the
MAGNETISM OF MARS?

Since this question is still wide open, I don't have to step on any big toes here. The latest theory (2007) is that Mars' magnetism was blasted away by asteroids, but that is so desperate I won't even comment on it. And I won't offend many by not taking it seriously. Mainstream physics doesn't have a charge field at the macrolevel, so it simply can't answer questions like this in a reasonable manner. It is at its worst trying to answer these sorts of questions, and seems to recognize that, so it rarely even tries to. Only physicists desperate for attention publish theories like this asteroid theory, and these theories tend to have a shelf life of about six months. They are fodder for the covers of the science rags, since they lend themselves to glossy illustrations, and then they die.

But this is an interesting question for me, especially. I have already explained the lack of magnetism on Venus as due to the fact that it is upside down. When its magnetic field is emitted from the surface, this field hits the ambient field. Since one field is upside down to the other, they cancel as a matter of spin. Yes, I have shown that magnetism is a function of photon spin, and the photons coming out of Venus are upside down relative to the photons not coming out Venus. Compared to the Solar system field, Venus is emitting anti-photons. We have a spin cancellation. It is that simple.

With the Moon, we have a slightly different mechanism. As the mainstream tries to explain Venus' lack of magnetism by the slow rotation of Venus about its axis, they do the same with the Moon. But this can't be the cause, because both Venus and the Moon have strong electrical fields. Venus, especially, has a powerful ionosphere, one that blocks the Solar Wind much like our magnetosphere. If the lack of rotation damped the magnetic field, it would damp the electrical field as well, and we don't see this with Venus. So something else must be happening there, and on the Moon. Again, we have to study the direction of the spins of the photons being emitted out of the Moon, since it is these spins that cause the magnetism. The ambient field (the field around the Moon) isn't upside down relative to the Moon, but it is opposite in spin in another way. Since the Moon is so close to the Earth, the Moon's ambient field is determined more by the Earth than the Sun. This is not the case with Venus, obviously. So the field emitted by the Moon is always meeting the field emitted by the Earth head-on (at least on the nearside). Since the linear vectors are opposite, we again get a spin cancellation.

You will say, "But this means we should find more magnetism on the far side!" Well, in fact, we do. As just two examples, I point you to Mare Ingenii and the Gerasimovich crater. We have been informed by NASA and Russia of "magnetic anomalies" in such places, but NASA will not admit that these findings are not that anomalistic. What they mean by anomalistic is "it goes against our assumptions." But an anomaly is usually something that contradicts other *data*. And we don't have a lot of standing data indicating the farside of the Moon has or should have an absent magnetism. More cratering is often used as evidence of that, but that is evidence of nothing. The Moon has more cratering over there simply because it isn't protected by the Earth over there nearly as much, either the body of the Earth or the field of the Earth. The Moon's ass is out in the wind, if you like, and so of course it will show more cratering.

But my theory wouldn't predict that much more magnetism over there anyway. Why? Because the Earth's photons are blocked from the farside. It is Earth-dark over there. So the photons coming out of the Moon don't feel much spin boost from the Earth's photons (except those going through the Moon). The Earth's photons only set up a sort of wall, keeping the Sun's photons from defining the ambient field, and therefore providing the boost themselves. The Moon's photons on the farside are emitted into a flat field, as it were. Their spin comes only from the Moon's interior magnetism.

So although they aren't damped, they aren't spun much either. They therefore have a low magnetism. Higher than the nearside, but still low.

Notice that this also explains local fluctuations on the Moon, such as at Gerasimovich crater. That is where the Earth's photons are going through the Moon. We are seeing the result of density variations and material variations inside the Moon. On certain trajectories through the Moon, more photons get through to the farside. When they come out over there, we see magnetic maxima.

But with Mars, I have neither of these answers to use, do I? Mars is not upside down, like Venus, and it is not spinning that slowly, and it is not in the shadow of some other very near body. Its day is about the same as the Earth's, and its radius is about half, so it seems at a glance that by my theory it would have about half the charge and therefore half the magnetism. But that "at a glance" is way off, since we have left Jupiter out of it. We can't do that, as I showed in both my axial tilt papers and my Bode series paper. Charge moving toward the Sun increases in charge density and therefore in charge power, which also increases the magnetic power. This is just to say that the photons get closer together, because they are moving into a smaller volume. Therefore, Mars IS is the shadow of another body. This despite the fact that Jupiter is so far away.

Let's use the math I used in those earlier papers to show this. The charge density of Jupiter is 1/986 that of the Sun, and Jupiter is 2.44 times further away from Mars than the Sun is. The fourth root of 986 is 5.6, so the relative strength of the Sun's field is 5.6. The relative strength from Jupiter is 2.44.

So the total relative strength of the ambient magnetic field at Mars due to Sun and Jupiter combined is 3.16. But that is still disregarding the other big planets. Saturn is 1/7.28 relative to Jupiter, and 2.19 times further away, so its charge at Mars, as a fraction of Jupiter's, is .3. Which gives us .3 x 2.44=.734 as the input from Saturn. Uranus is 1/25.8 relative to Jupiter, and 4.81 times further away, so its charge at Mars is .186 x 2.44=.455. Neptune is 1/16.9 relative to Jupiter and 7.77 times further away, so its charge at Mars is .46 x 2.44=1.12. Add them all up, and we get 4.75. That means we have 5.6 from inside Mars and 4.75 from outside (disregarding the smaller planets). The difference is .85.

But what does that mean? Well, it is a relative number, not an absolute number, so we still have to compare it to the Earth by the same method. From those other papers, we know the total charge *density* from the outer planets is 5.66 times that from the Sun, at the Earth. So if the Sun's charge is 5.6, their total charge is 31.7. This means that the ambient field is 30.7 times stronger at the Earth than at Mars. Or, at Mars there is much more balance from within and without, so that the field differential is much less. If we study how this affects magnetism, we find it means the photons at the Earth are spun almost 31 times as much as the photons at Mars.

So we multiply that effect by the other effect. I said the "at-a-glance" number was 2 above, but that is not correct either. It isn't just a matter of radius. We can see that just from the torque equation ($T=Fr$) or the angular momentum equation ($L=mvr$). In other words, we have to use a mass *and* a radius, not just a radius. Even according to current math, that would be true. But I have shown that charge follows both mass and density, not just mass, so again we find a charge density or mass density. Mars is .0763 that of the Earth, which we then multiply by the radius differential of . 533, which equals .04. We then multiply that by the number we found above, 1/30.7, which gives us .0013. That would be my rough estimate for the magnetism of Mars: about a thousand times less than the Earth. That matches current estimates and data, which run from 10^{-3} to 10^{-4}.

One last thing to hit, before I finish. We know that although Mars has a low current magnetism, some of the rocks on Mars have a much higher residual magnetism. This has been taken to mean that Mars had more magnetism in the past. I think this is entirely possible, and that this reading is probably correct. But I do not think the magnetism was knocked off the planet by asteroids. No, this residual magnetism in the rocks on Mars is telling us something very important, not about Mars, but about the make-up of the Solar system in the past. It means that either Mars was not at its current orbital distance at that time, or the big outer planets were not. Something was vastly different. Given the asteroid belt and other glaring evidence, this is not hard to imagine.

From this we see that my theory of planetary magnetism will give us the tool to work backward in time, rebuilding previous Solar system relationships. These rocks, which we find on other planets and moons as well, are like tablets with numbers on them. They will be very useful in future.

Chapter 12

C-ORBIT ASTEROIDS

can only be explained
by the Unified Field

Asteroid 2010 SO16, discovered in September of 2010 by Apostolos Christou and David Asher at the Armagh Observatory in Northern Ireland*, follows a C-orbit or horseshoe orbit which approaches the Earth and turns around. Of course, by the gravity-only theory, this is impossible. To divert you from this rather obvious and glaring conclusion, the mathematicians at NASA and worldwide shunt you off into Lagrange point math. I have already shown my readers how they fudge these Lagrange point equations in a previous paper, but I can show you why it is impossible very quickly.

The basic equations of gravity were invented by Newton, of course, and they have never been overwritten. They have been updated with time differentials by Einstein, but Newton's equations stand beneath the field equations. Certainly neither Laplace, Lagrange, nor Einstein ever denied that gravity increases with decreasing distance. None of them falsified the inverse square law. It is still taught in all physics books, and it is bedrock to this day. That being so, all we have to do is look at the asteroid as it approaches the Earth, from either direction. The distance between asteroid and Earth is diminishing with time, so the gravitational force between them must be increasing rapidly. The distance between Sun and asteroid is not yet changing at these points (it changes soon afterwards), so the force between Sun and asteroid is not changing. Therefore, we may ask what would make the asteroid make a 90° turn at this point in its motion. Even more to the point, what would make it turn another 90° and move away from the Earth? We need a force or other mechanical cause here, not just math or field lines. Neither math nor field lines can turn an asteroid. The force of gravity, which is supposed to be beneath these field lines, defining them, cannot possibly cause the asteroid to turn around and move away. Gravity is a force of attraction, remember? And we have a diminishing distance here, which should cause a steady increase in attraction. By all the laws of gravity, the asteroid should crash into the Earth. It doesn't, so the gravity-only theory cannot be correct. And no amount of pushed equations can save the gravity-only theory, or convince us that gravity can *repel* an incoming asteroid.

As I often say, it is beyond belief that I have to be on this page saying this. It would be like a basketball critic needing to tell other professional basketball players that when you throw the ball, you should throw it at the basket instead of into the stands.

My charge field solves this problem in just as spectacular a fashion as the current theory fails to solve it. It solves it with the Unified Field. The second half of the Unified Field is the charge field, and the charge field is repulsive. Charge is just emitted photons, and they work by straight bombardment. Just as the solo gravity field increases with decreasing distance, so does the charge field. But the charge field increases even faster than the gravity field. Gravity increases its attraction by the square, but charge increases its repulsion by the quad. So the charge field is capable of bouncing out intruders, even while the gravity field is still working full strength.

How does it do this? It does it with simple mechanics. The nearer you get to the surface of a sphere, the denser the emitted charge field becomes. And this is due simply to the surface area equation. You have the same field in a smaller space, so the density rises.

This is not some wild hypothesis on my part. I have simply brought the charge field that is admitted to exist at the quantum level and applied it at the macro-level. A force cannot appear at the quantum level and disappear at the macro-level. Current phycsis tries to dodge this question by making charge at the quantum level virtual, but that is not physics. "Virtual" is the opposite of "physical", and both the math and the "physics" of virtuality should be outlawed. Furthermore, I have shown where this charge field fits into Newton's equations, Coulomb's equation, and the Lagrangian. I have not just cobbled together a theory, I have done all the math. You will say they have done the math, too; but my math and theory are mechanical at all points, theirs are not. My math and theory is also simpler and far more transparent, since I take the time to label all the variables and explain all the motions. They never do this.

Looking at any C-orbit diagram, any teenager can see the repulsion. The asteroid is being turned by a field. The E/M field of the Earth is excluding the asteroid. I will be asked why it excludes this asteroid, but doesn't exclude smaller bodies, like meteorites. Again, it is strictly a matter of E/M field interaction. This asteroid has a diameter of several hundred meters, so it has not only an appreciable cross section, it has an appreciable E/M field of its own. A high density may add to this field. A low incoming velocity also helps. The asteroid is traveling about 175 times *slower* than the Earth, as both objects move around the Sun. This means that as they approach, the velocity belongs almost entirely to the Earth. The Earth is moving about 29.78km/s, so the asteroid is moving only .17km/s. And the combined velocity is then 29.95km/s. This means that the Earth is moving at the asteroid only .0057 faster than it is moving at the Moon. We shouldn't be so surprised that the E/M field of the Earth could counter such a motion, since we see the Moon repelled more than that everyday. Meteorites pierce this field because they are smaller and are moving much much faster. Meteorites reach speeds of 100km/s, which is almost 600 times faster than this large asteroid.

*http://earthsky.org/space/asteroid-2010-so16-is-following-earth-in-its-orbit-around-sun

Chapter 13

The

CORIOLIS EFFECT

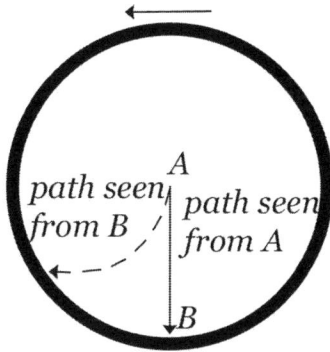

This is another phenomenon that has been badly misconstrued. We are told that bathtubs drain one way north of the equator and the other way south, and that weather patterns do, too, whether in water or air. I am not questioning the data, I am questioning the explanation, which is woeful in its lack of power and logic.

It is admitted that the Coriolis Effect is not a real force. It is only an outcome of circular motion. A line that looks straight from one position will look curved from another position. Again, I am not doubting that. I accept both the data and much of the math. However, I think it is clear that the Coriolis Effect is only an effect of pre-Einstein relativity.

That is, it is an outcome of position and motion, not of forces or dynamics.

Many physicists will agree with that, but I will go further. When it is used to explain vortices on the Earth, it is false. It cannot logically explain them. To explain these vortices, we require the charge field.

Some will stop me here before I get started, telling me that I don't need to go to the trouble. We can solve this very simply without either a long-winded mathematical analysis of the Coriolis Effect or of the charge field. At a website called "Bad Coriolis", the author, while critiquing some of the current uses of the Coriolis motion, simplifies the argument into this: The Earth is spinning counter-clockwise in the northern hemisphere, and so are the hurricanes: nuff said. While I admit that explanation is preferable to the current mainstream one in many ways, it still begs the big question: WHY is the Earth spinning counter-clockwise, or to the east? Why not to the west? As it happens, his explanation is only partially correct. This is because the charge field underlies the spin of the Earth itself, as well as the spins of hurricanes and so on. As I will show below (and as I have already shown in previous papers), the ambient or Solar charge field determines all the local fields in the Solar System, and by doing so, determines the spin direction of all the planets and moons. I have shown how it causes tilt, eccentricity, and other variables, and here you will begin to better understand how it causes spin. If you have been following the titles of my papers this past twelvemonth, you will have seen that the charge field causes almost everything.

So vortices on the Earth cannot be caused by the Coriolis

Effect alone. The easiest proof of that is this: if the vortices were caused by the Coriolis Effect, and switched at the equator, then there should be some point on the equator where water drained with little or no spin. We have never found that place, therefore the theory is falsified. The theorists are required to explain the negative data, and they cannot do it.

The simplest way to visualize this is to imagine a merry-go-round or carousel spinning in a zero gravity field. With no gravity, we could put polehorses both on top and below the spinning carousel. The children could spin upside down or rightside up. They could just crawl under the carousel and spin on a second ride. This is the way physicists now imagine and explain the Coriolis Effect. If the children on top see the ride moving clockwise, the children on bottom see the ride moving counterclockwise. Drain problem solved. The children also sees curves from center to edge reversed. Large weather curves solved.

I admit that is somewhat ingenious, which is why I accepted it for years. But we encounter big problems if we let a child stand on the edge of the carousel, right on the equator. He doesn't see any curve at all, or at least not in the same plane as the other children. If two children stand on the edge and throw a ball to one another, the second child will see the ball mysteriously rise (the Eötvös effect). The curve will be up. But the Coriolis effect proper is gone. Do we find this in studying drains on the Earth's equator? No. Do bathtubs drain up or fail to drain on the equator? No. Do they drain without spin. No.

The Coriolis Effect also fails to explain the tight curve of drains and cyclones and so on. The children on the merry-go-round see curves that correspond to the curvature and speed of the ride. They do not and could not possibly see little vortices at spots on the ride. Nor would making the ride into a sphere rather than a circle create these little vortices. Yes, we find these little vortices on the Earth, but they cannot be caused by gravity, the Coriolis effect, centrifugal forces, or all three combined. Inertial circles, as they are called, cannot be the outcome of inertia, of any of these forces or pseudoforces.

Another big problem can be seen by studying the figure under title. The view from A is the view from the center or pole. Well, we can go to either pole of the Earth and look at weather from there. And we can take planes and helicopters well above the poles, or put cameras in high flying balloons. Do they see curves in weather straighten out? Do they see cyclones and hurricanes stop spinning? No, these curves are real curves whose curves do not depend on your perspective. The various vortices in weather and drains are not caused by relativity or by position or by pseudo-forces like the Coriolis Effect. They are caused by something else entirely.

Here's an even bigger problem. The Coriolis Effect is used to explain the deflection of a cannonball in various thought problems.

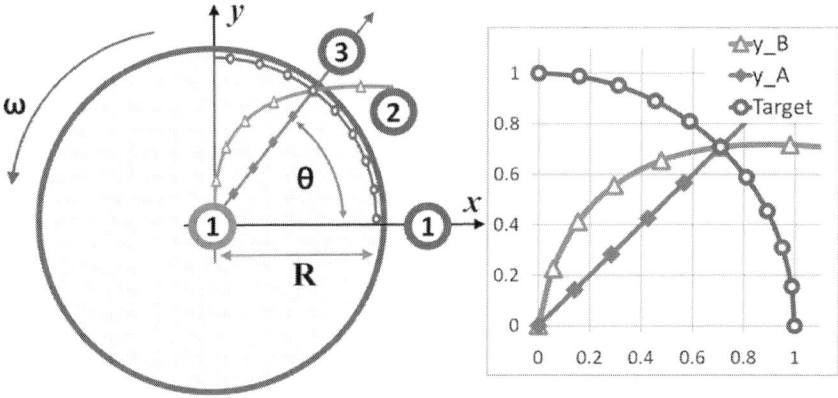

Unfortunately, that illustration contradicts the figure below title, since we can see there is no real deflection. In other words, it takes no force to deflect the cannonball, since it is not really deflected. It only *appears* to be deflected due to the position of measurement. If we are looking at the cannonball from off the turntable, we won't see the Coriolis curve; and yet in this illustration we are off the turntable and we do see it. The illustration is falsified. The authors admit this, yes, but the visuals are still confusing. We aim ahead of the target not because the cannonball curves, but because the target moves toward the line. The cannonball is not accelerating, the target is. And there is no gain in energy from the curve either, since no force was used and no acceleration was present. Again, the acceleration is only apparent, due to position of measurement. We think a curve must require an acceleration, but in this case it doesn't. The force is pseudo so the acceleration must be, too. Well, that is a problem for hurricanes, since hurricanes don't have pseudo-energy. They have a real energy gain from the vortex. That being so, the cause of the hurricane cannot be

179

the Coriolis Effect.

Yes, the spin of the Earth creates weather patterns. It creates latitudinal currents which, when they meet longitudinal currents, create curves and vortices. I am not denying it. I am not here to analyze or critique all of meteorology. I am only pointing out that the longitudinal currents, when curved by Coriolis Effects alone, cannot have any real power beyond their straight-line velocity (or their centrifugal power). They cannot be the cause of the tight curve even in the largest hurricane, because Coriolis curves don't curve that much. And they cannot be the cause of the energy of the hurricane, because Coriolis curves don't have any real energy. The curve of the cannonball in the illustration can't have any more energy than the straight line in the figure above, since they are the same.

We can see this again by looking more closely at a hurricane. This "low pressure system" is over Iceland.

Notice that we have more than a Coriolis curve here. We have three or four complete circles. Why does that matter? Because when the curve is moving up from lower latitudes to higher, it is actually moving *against* the spin of the Earth. To put it another way, it is anti-centrifugal. A real Coriolis curve always moves out from the center or the pole, since that is the "force" of the spin. Put a marble on a record player near the center hole and then let it go. It moves out. If you put a marble on the outer edge, it will fall off. It will never move toward the center. Yes, if you push it hard, it will go to the center, and will create a Coriolis curve in reverse. But you must push it. That push is a real force. You have to counter the centrifugal force or motion. Some force is counteracting the centrifugal motion of the Earth in this hurricane, and it isn't the Coriolis force. We are told that the centrifugal force of the Earth isn't that high, but that is false. In this case, it is very high. The Earth has a lot of angular momentum, and to get anti-centrifugal motion on this scale and at this speed requires real forces, not pseudo-forces.

We see the same problem when we look at the motion of the hurricane latitudinally, or parallel to the equator. Neither centrifugal motion nor Coriolis motion can move that way. Centrifugal motion is always away from the pole, and Coriolis motion is, too. Anytime the air in the hurricane is not moving away from the pole, we require another explanation for both its motion and its curve. With centrifugal and Coriolis motions, we can explain motion south (in the northern hemisphere) and west, but we cannot explain motions north and east. Just consult the animation under title once more. The disk is spinning east, like the Earth, and the ball is curving west.

181

And this brings us to the killer punch. Look again at our hurricane over Iceland. Now look at the figure below title. Now look at the last sentence of my last paragraph. Do you see a contradiction? The hurricane is backward. The Coriolis Effect should cause a curve east to west, as you go south. The hurricane is spinning the other direction! This is a picture of a hurricane that is anti-Coriolis and anti-centrifugal. The subtext says it "spins counter-clockwise due to balance between the Coriolis force and the pressure gradient force." False. The Coriolis force is in the other direction, so it cannot balance any pressure gradient like this, no matter where it is coming from. We are told that "Low pressure systems rotate in the opposite direction, so that the Coriolis force is directed radially outward and nearly balances an inwardly radial pressure gradient." Criminy, these people are shameless. They expect you to believe that! Just look at their own diagram for this:

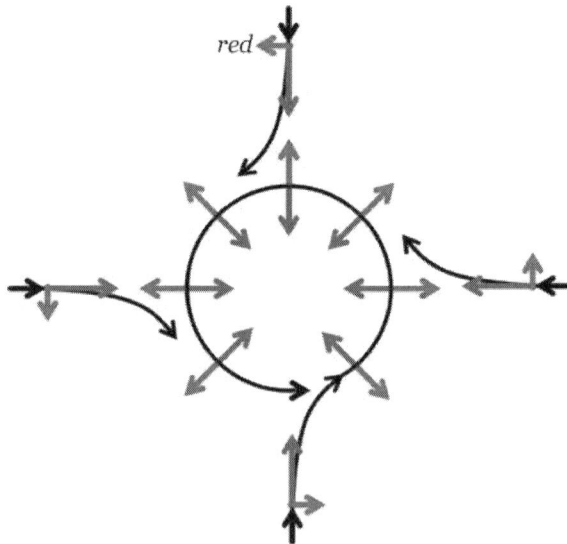

Study the red arrows, which we are told represent the Coriolis acceleration [the red arrows are the shortest ones on the outside, pointing counter-clockwise]. Notice that we have red arrows pointing north and east. Impossible. The Coriolis motion cannot be north or east. Period. Ever. These diagrams are simply matched to the data, and then the geophysicists or meteorologists just attach whatever tags they like to the vectors, with no concern for whether they make sense or not. They figure no one is going to study this stuff closely, so why bother making sense.

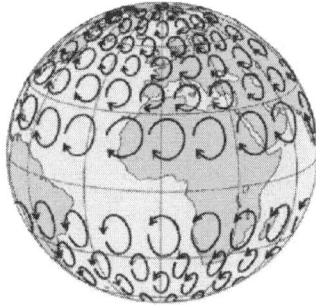

Inertial circles also cannot be explained by the Coriolis effect, for the same reason, and this dooms all of current meteorology. You need inertial circles to explain low pressure circles, according to the current math and diagrams, so if inertial circles are a fudge, the whole thing is a fudge. In the northern hemisphere, only the south and west motion of the circle can be attributed to the Coriolis effect. But since there is no possible north and east motion, we cannot complete the circle. The Coriolis effect might be able to create half circles, but it cannot create full circles. I draw your attention to this quote from Wiki:

An air or water mass moving with speed *v* subject only to the Coriolis force travels in a circular trajectory called an 'inertial circle'.

Subject only to the Coriolis force. They just said it themselves. They are not creating these circles with any other force or motion. Impossible. These circles are also too small to be curves caused by the spin of the Earth. The Coriolis motion doesn't work that way. Again, they are just matching the diagram to data. They know that circles this size are needed to explain the low pressure systems they see, so they create them in the math. This is the math used:

$R = v/2\pi f$

Where R is the radius of the circle and *f* varies with latitude. Unfortunately, that math is pushed as well, since on a spinning planet where gravity was the only other force, you couldn't get *f* to give you these small circles. This is because gravity doesn't vary over the surface, so it can't give you a variation with latitude. And the Coriolis effect can't either. The Coriolis motion can only give you a greater curve as you get closer to the equator, but it can't give you multiple curves. The fact that the Earth is a sphere rather than a circle isn't enough to create these breaks at latitude, where the Coriolis motion becomes flat and then begins curving back up. The math is pushed to match weather data, as I said, but it doesn't match pure physics.

You will say, "You have admitted that the Coriolis motion creates a curve. Does the 'force' really have to apply all the way round the circle? Can't it just push for part of the circle? If you push someone on a circular swing, you don't have to push all the way round. You just push once for each rotation, right?" Yes, but that example is not analogous to this

184

problem, since in a circular swing the swing is tied to the center. We don't have any such constraint here. If the rotation were already defined, then one push could keep it going, but physicists are using the Coriolis force to define the circle itself. That can't work with one push, or even a push during half the circle.

The biggest problem is the small size of the inertial circles they are trying to create. You see, the curvature of those circles is much greater than the curvature of the Coriolis curve at that latitude. The Coriolis curve is really just one big curve running from pole to equator (center to edge, same thing), as in the figure under title. The curvature at a given latitude is defined by that one curve, and it can't be any other curve. Nor does it matter where you start. If you start at 60 degrees north, for instance, and let the Earth spin 10 degrees east, the Coriolis curve will move an object 10 degrees west and some smaller amount south. So what it really creates for an observer on the Earth is a spiral. But the observer can't even observe the spiral, since the sphere will be blocking his view most of the time. The observer won't see inertial circles, he will see the object in Coriolis motion move pretty much directly away from him to the west and then disappear over the horizon. About 23 hours later or so the observer will see the object come over the eastern horizon, fractionally further south than it was before. So the only circles the object is creating are latitudinal circles, and that only because the Earth is creating them. Neither the Coriolis motion nor the centrifugal motion is really creating the circles. The centrifugal motion is due south, with a curvature that matches the curvature of the Earth; the curve of the Coriolis motion is measured in how much the spiral increases each day. So the only circle is the circle that

185

motion makes around the Earth each day. The observer could not possibly see that circle as an inertial circle, since nothing he sees ever goes north or east. In fact, he couldn't see it as a circle at all, since he only sees the object when it is passing him by in a nearly straight line. I hope you can see that no hurricane could ever hope to be created that way.

What no one seems to understand on these pages is that the Coriolis force is mechanically linked to the centrifugal force. You can't have a Coriolis force without a centrifugal force, and they are tied to eachother at all times. This is because they are both outcomes of spin. Therefore, the Coriolis motion is always going to be to the south in the northern hemisphere, because that is the direction of the centrifugal force. The Coriolis motion can never have a northern component, because if it did it would be anti-centrifugal. If it were anti-centrifugal, it would be anti-spin. The Coriolis motion cannot be anti-spin. That would be like weight being anti-mass. It conflicts with the definitions of the words. The same applies to an eastern component. There is no possible eastern component to the Coriolis force, by definition. This means that no observer can possibly see the Coriolis motion make a circle, except a latitudinal circle around the Earth over the span of 24 hours. If the Coriolis motion is never moving north, no possible observer can see it move north. The only observer that could see a Coriolis motion move north is an observer moving south, without knowing it. But that is not the case here. We do not have ignorant south-moving observers cataloging hurricanes, with hurricanes invisible to everyone else.

Another huge problem is encountered when we look at friction. The Coriolis curve can only be caused when the object making the curve has no friction. That is why "frictionless" or very low friction turntables are used when showing the effect at small scales. The reason we need no friction is that the curve is caused by the difference between an observer on the turntable moving with it (WITH friction), and an observed object moving without friction. The difference between no friction and friction causes the appearance of the curve. The observer spins and the observed object does not. Therefore, if the observer defines himself as motionless, he will see the object appear to curve. That is what the Coriolis motion is. But this means that whatever is claimed to be in Coriolis motion on the Earth should be frictionless or of very low friction. That isn't what we find. Water and air have lower friction than solids, but they are far from frictionless. We already know that both air and water are carried along to a large degree by the spin of the Earth, for if they weren't it would be quite obvious. The oceans would swamp all the Eastern shores, and the atmosphere would move to the west at a constant and high velocity. On the equator, the wind would always be blowing 1670 km/hr, which would be pretty hard to miss. It is true that friction isn't the only thing that prevents this, but it doesn't matter here. What matters is that the air and water are NOT moving like a frictionless ball moves south on a turntable. The air and water are moving with the Earth to a large degree, which means they are moving along with us spinning observers, which means we observers would not be expected to see much of a Coriolis effect. To the degree that the water and air spin with the Earth, the Coriolis effect is nullified. If the air is mostly moving along to the east with you, you cannot see it move to the west, can you?

To see how confused contemporary physics is once more,

just look at the math they have included for Coriolis effect. This is the diagram:

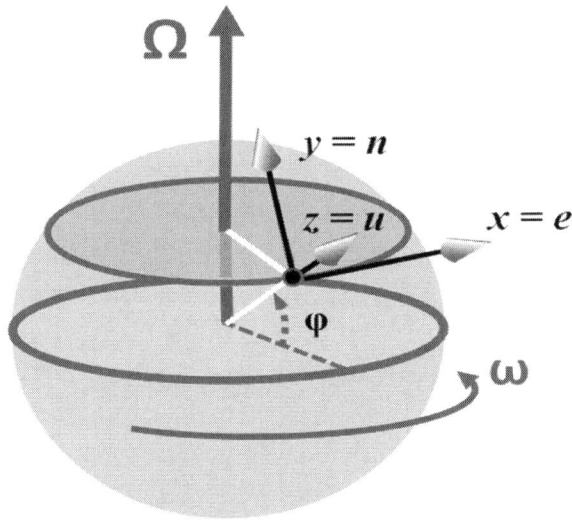

As you see, they have their planet spinning east, and they are diagramming a point in the northern hemisphere. But they have the Coriolis force divided into vectors north and east! That is upside down. The Coriolis motion in the northern hemisphere is south and west. Actually, what they do is even nuttier than that. These vectors they have drawn are not Coriolis motions or forces at all, they are just samples of positive motion. They are letting east equal +x and north equal +y. Then they find that motion east creates an acceleration to the south and motion north creates an acceleration east. If you aren't confused by that, they haven't done their job, for the whole point of this math is to make your head spin. If you are dizzy enough, you will accept anything they say.

Ask yourself this: Why don't they just solve for a particle

placed at that point, instead of creating these stupid initial motions? Because if they did that, you would discover that the Coriolis acceleration on that particle was south and west, the opposite of their drawn vectors. You would understand what the Coriolis force really was, and then all their other diagrams and explanations would begin to crumble in your mind. You see, this math and diagram are misdirections. They not only very cleverly hide from you the fact that the Coriolis motion must be south and west, they actually fool you into thinking it is or might be north and east. Most people won't pull apart the math like I did, they will just look at the drawn vectors, and they will think that the Coriolis acceleration can be north or east. If they think that, they won't question the other diagrams or math.

Again, consult the figure under title. The Coriolis motion is away from the center of the circle or away from the pole. That is the centrifugal part of the motion. The curve is opposite the direction of motion of the spin. So if the Earth is spinning east, the motion must be west. The Coriolis curve must be south and west in the northern hemisphere. It can't be anything *but* south and west, and it can't create little circles to suit these people.

To counter this, we are shown circles created on a parabolic turntable, as if that is to the point. It *isn't* to the point, since the Earth is not a parabolic turntable. But again, I don't have to do any math, I only have to point out that the Earth cannot be analogous to the parabolic turntable because we can now get off the Earth quite easily. We can look at the Earth from an inertial frame of reference just by hovering over one of the poles, and when we do that neither the inertial circles nor the opposite spinning weather supposedly created by them revert to straight lines. At Wiki we get very little on the parabolic turntable, but you can go here* to see how it works. To the scientist watching the turntable, the circles don't appear. You would have to go onto the turntable to see the circles. We know they are there by using a camera above the turntable, rotating with it at the same speed. Playing back the film, we see the inertial circles. But two things may be said against this, 1) I repeat that extending the poles of the Earth creates an inertial frame relative to the Earth. If you are off the Earth watching the Earth spin, you are like the scientist off the turntable watching it spin. You shouldn't see the circles or the weather they create. However, you do see the weather, therefore the theory is false. The circles aren't created that way. 2) Even if I can't convince you of that—because you believe (wrongly) that Einstein proved that no frames of reference are inertial—you should see that these inertial circles on the parabolic turntable aren't analogous to any possible motion on the Earth's surface, simply because they are caused by simple harmonic motion. You see, to create the circles, the physicists had to create harmonic motion. They needed a closed circuit, and that is what harmonic motion is. The ball on the parabolic turntable goes up and back, so it creates a closed circuit, both in the inertial frame and the non-inertial frame. But the Earth's surface isn't like that. They tell us that the Coriolis curve is parabolic in that the curvature increases with distance from the center or pole, but that is the curve of the moving body, not the curve of the Earth. The curve of the Earth is not parabolic, any more than the curve of a normal record player is parabolic. But if they were going to turn the flat record player into a parabola, to match it better to the math somehow, they should have built a convex parabola, not a concave parabola. The concave parabola, with the center lower than the edges, creates harmonic motion and a closed circuit. But the convex parabola can't do that, for obvious reasons. The object accelerates to the edge and then flies off. The thing is, the Earth is analogous to the convex parabola, not the concave parabola. We can see that just by looking at where the greatest velocities are. With the concave parabola, the greatest velocities are at the center. With the convex parabola, the greatest velocities are near the edges. The Earth is obviously a convex parabola, in that sense, in that the greatest velocities are near the equator. The least velocities are near the poles. And that is true both of velocities caused by spin and velocities due to the Coriolis motion. Therefore, the real Coriolis motion on the Earth cannot create a closed circuit. It gains velocity as it goes south until it reaches the equator, and then the acceleration stops (because the curvature of the Coriolis curve stops curving). The particle does not fly off the Earth, as it would the convex parabola, but it stays at the equator. It does not curve back up, because nothing is compelling it to do so. All this is very clear I hope, so you should see that the deflection into a concave parabola, and all the math included in that, is just another hoax.

Wikipedia addresses this in only sentence:

On a rotating planet, f varies with latitude and the paths of particles do not form exact circles. Since the parameter f varies as the sine of the latitude, the radius of the oscillations associated with a given speed are smallest at the poles and increase toward the equator.

Funny that these authors of a science information site have the time to include the math for the rotating sphere, the fictitious force, the Rossby number, the flight of the cannonball, the tossed ball, and the bounced ball, and have

191

time to mention the Eötvös Effect, the parabolic turntable, ballistic missiles, and molecular physics, but do not have time to give us more than one sentence on this. All the other math and physics depends on this, but this is hidden from sight! Do you not find that the least bit strange? Well, I have shown it was not an oversight. This is all they have to say, because this is false. "They do not form exact circles" is hedging in the extreme, since I have shown that they do not form circles at all. And "the radius of the oscillations...are smallest at the poles and increase toward the equator" is also misleading, since, although it is true, it applies to one big curve, not a lot of isolated ones. If you do the math on the Earth, instead of on the parabolic turntable, you get one big Coriolis curve and no little inertial circles. That is precisely why they divert you off into the parabolic turntable. If the math and diagrams for the Earth showed you what they wanted you to believe, they would have just shown you that, right? Ask yourself why you need to be shown the parabolic turntable, when you can just as easily be shown the Earth. Instead, they have a section on the parabolic turntable, and no section on how the inertial circles are created on a sphere.

It is interesting to note that tornadoes are not explained by the Coriolis force, since it seems clear that such a small tight curve cannot be explained that way. Wikipedia says, "while tornado-associated centrifugal forces are quite substantial, Coriolis forces associated with tornadoes are for practical purposes negligible." But this doesn't prevent even tighter curves like drains from being explained by the Coriolis force. We don't get a Rossby number for drains, we just get some bad and limited experiments and the assurance that it must be the Coriolis force once again.

And this brings us back to the drain problem. Notice that drains in the northern hemisphere drain counter-clockwise, like the hurricane but *not* like the Coriolis motion. We should find that curious, because we now need a lot of low pressure system, gradient force gobbledygook to switch the direction there, too. We need a lot of very tiny inertial circles in your bathtub, surrounding areas of low pressure, like little gears and cogs. This would act to switch the clockwise Coriolis motion to the counter-clockwise drain motion.

Yes, I have uncovered another big farce. Meteorology and geophysics contain some good math and good models. They also contains a lot of very bad math and very bad models, as we have seen. The problem is that the theory under these vortex models, like the theory of tides, conceals a big hole. In the case of large weather patterns, we have the Coriolis Effect substituted for the charge field. Current physics doesn't have the charge field to work with, so it has to fill that hole somehow. In celestial mechanics, it fills that hole with Lagrangians and other fancy math. In these curves in wind and weather and water, it fills the hole with the Coriolis Effect. Likewise with many smaller effects, like the vortex of a drain. Without the charge field, physicists can only fall back on the Coriolis Effect. But I have shown that it doesn't work.

Notice that we would *expect* the charge field to act differently north and south, since the Earth is a sort of dipole. I have denied that charge is dipole by the old definitions, but I have not denied that the Earth acts as a dipole, with different charge motions at one pole than the other. These motions are not caused by repulsions and attractions, but they are real. In a nutshell, the spin of the Earth causes low charge pressure at the poles (by mechanical means only: see my other papers, most recently the ice age paper), which causes an intake of charge photons at the poles. But one pole intakes photons and the other intakes anti-photons. The Earth then recycles this charge, flinging it off most heavily at the equator (due simply to angular momentum peaks there). Although charge is heaviest at the equators, it is emitted everywhere. The photons and anti-photons remain sorted, however, with one being emitted more heavily north and the other being emitted more heavily south. Again, this sorting is done strictly mechanically, with the division being caused by their initial velocities into the poles. They are diverted by existing charge fields in the Earth (electric and magnetic), but since they are coming from different directions, they are diverted in different arcs. This is what causes the split to remain split.

Now, the difference between photons and anti-photons is only a difference of spin. One is upside-down to the other. And it is this spin of the photons that causes magnetism, as I have shown in detail elsewhere. So the fact that you have more photons in one half and more anti-photons in the other means that the magnetism north and south will be reversed.

We already know that, in part. We don't know the cause, but we know the effect. We know the magnetism is reversed north and south, since that is what we mean by north and south poles. If the magnetism weren't reversed, both poles would be north, and a compass could point to either one, depending on your latitude. But the magnetism isn't just reversed at the poles. It is split from the equator out.

Because the charge field is different north and south, we would expect vortices to be different north and south. And this would apply to vortices of every size, large and small. Since the curves are not caused by position or by Coriolis pseudo-forces, there is no need to explain the vortices by lots of difficult math. Magnetism is caused by a real spin of a photon, so we can explain angular momenta all the way down to the size of a photon. Small vortices give us no theoretical problem. And larger vortices are just collections of smaller ones. Because the charge field is ubiquitous and quite strong everywhere on the Earth, every point on the Earth will have a predisposition to vortex one way or the other, depending on the phenomenon. But because friction and gravity are even stronger, these predispositions show themselves only in limited circumstances. They would be most likely to show themselves in liquids and gasses, of course, where friction is limited. And they would be most likely to show themselves in the presence of ions, the heavier the better. This is why they show themselves in weather: storms are strongly ionic. Water is also known to be a good conductor, especially salt or mineral water, so it is no surprise to find these vortices in water.

I would say that the interesting experiments have not yet been done in regards to this phenomenon. Vortex experiments should be done at equator and pole, and compared, not only for direction but speed. Then magnetic fields should be applied, to see how these affect the speed at both places. Then ions should be introduced in varying amounts, to see how this affects the speed of the vortex. Various liquids should be introduced as media for the vortices, using liquids of high and low conductivity. I expect the liquids with higher conductivity would create quicker vortices. Any or all of these experiments would immediately doom the Coriolis explanation, since the Coriolis Effect could not possibly be increased or decreased with ions, magnetism, or varying amounts of conductivity.

In closing, let us look at the actual curves. Current theory is forced to do a complete switch, since the Coriolis force would show hurricanes spinning clockwise and they actually spin counter-clockwise. Same with drains. This shows that data is never a very high wall to climb, given the right math. These mathematicians can turn night into day when they like. But with charge, we don't need to do that. Our charge photons are already spinning counter-clockwise in the northern hemisphere, so we don't need to finesse pressure gradients to explain hurricanes. Anti-photons spin counter-clockwise, they come in at the south pole, and they are emitted more heavily at the equator and in the north. Since they are always present, they predispose the entire unified field to spin with them, in the right circumstances. In most cases, the predisposition is only potential, but given enough ions and a lack of friction, it can be expressed. Since the cause is the actual field particle itself, we can explain any size vortex, even molecular or atomic vortices. We would expect material vortices to have a lower limit above the size of the ions present, since matter is normally driven by ions. That is, charge is normally expressed in the baryonic field via electricity and magnetism, which require ions. But at smaller scales, we would expect vortices caused by the photons directly.

I have shown that current Coriolis theory can't explain draining at the equator, so I should have to answer it myself. If we have a switch from CW to CCW, we should still have a line where the switch is made, right? Not really. With the Coriolis explanation, there is no good way to explain draining near the equator, which is why it is the one question never asked or answered by the talking heads. Go ask it to your Google search engine. I did and I got nada. With Coriolis and Eotvos and the rest, we would expect drains to drain poorly on the equator and to have very little or no spin either way. With my theory, we can explain local variations somewhat more easily. Like this:

Some have thought that according to my theory of recycled charge, the equator must show more magnetism than other points on the Earth. Their reasoning was this: if more charge is being emitted there, and assuming no lack of ions there, we should see stronger magnetic fields. That is a logical conclusion, but it fails for this reason. We have more charge, yes, but we have both photons and anti-photons. Both are being emitted, and both are being emitted in large quantities. So unlike with Coriolis theory, the equator is not a zero line or a minimum line, it is a maximum line. It is the maximum for both photons and anti-photons. Well, since spins cancel, this would mean that the magnetism would cancel. So by this way of looking at it, we should have more electrical effects on the equator and *less* magnetism. In fact, the two mechanisms offset: high charge means more magnetism and the high presence of both spins damps this added magnetism back down to normal levels. So the equator is neither much more nor less magnetic than other points on the Earth. It has more charge, and so we would expect stronger electrical fields, but not magnetic fields. Since it is the magnetic fields that cause curves, we wouldn't expect drains to act differently on the equator.

You will say, "But if we have equal amounts of photons and anti-photons, the magnetism should be zero, right? That is how you are explaining the lack of magnetism of Venus and the Moon and so on, if I remember." That is right. So we don't have equal amounts of photons and anti-photons. The Earth is not recycling equal amounts of each. The strong magnetic fields here tell us that we have a predominance of one over the other, and that is because the ambient field from the galaxy and Sun is unbalanced. The Moon has low magnetism not so much from balance as from lack of spin. It takes spin to maintain magnetism, and the Moon mainly recycles charge from Earth and Sun with little spin. This means the Moon's field has strong charge but weak magnetism. As for Venus, it is the fact that she is upside down that causes the lack of magnetism. She recycles plenty of charge, it just gets canceled in terms of spin when it meets the ambient charge field.

But back to the Earth. This must imply that the northern hemisphere should have more charge. Do we have any indication of that? Yes, we have many hurricanes in the North Atlantic, and almost none in the South Atlantic. Local magnetic fields hit a minimum in the southern hemisphere, in South Africa and South America.** We have more storms to the north overall, and although this used to be explained due to less detection in the south, this is no longer true. With satellite coverage of the entire Earth, we have found that there is indeed more "weather" in the north. This has also been attributed to greater landmasses in the north, but it may be that both the greater landmasses and the greater weather is caused by the same thing: more charge. Just as the planets inhabit the plane of greatest charge in the Solar System, it is probable that the land inhabits the area of greatest charge on the Earth. I will give you more reasons for that as my papers on charge continue to unfold.

I have said in many papers that celestial bodies emit more charge near the equator and less near the poles. In previous papers I provided a link from NASA of actual footage• of the spinning Sun, and it is clear from a glance that more charge is being emitted near the Solar equator. Do we have any similar glaring evidence on the Earth? Yes. We have known since the 1960's that the ionosphere is considerably weaker near the poles. As just one example, we are told in a paper† by Grote Reber (a pioneer of radio astronomy) that

Since these long waves must get through the ionosphere, the best locations for observing will be where the electron density is lowest. Examination of a vast amount of ionospheric data disclosed that there are two bands of about 35° latitude radius centered on north and south magnetic magnetic poles that meet this requirement.

I have said that charge drives ions, and here we have direct and longstanding data that we have fewer electrons being driven near the poles. That is direct proof that we have less charge at the poles. The only way to deny it is to say that E/M isn't driven by charge. That would be novel, since all of QM and QED and QCD is based on the idea that E/M IS based on charge. I have never disagreed with mainstream theory in this, I have simply given charge a real presence, rather than a virtual presence. And I have given it a real presence at both the quantum and macro levels. We know from data of these charge holes at the poles, but they have never been explained. I have never seen an explanation attempted. But, as you see, it is a natural outcome of my theory of charge recycling. We have less charge at the poles because charge is coming in there, not being emitted there.

So we wouldn't expect ions to be driven up. They would be driven *down* at the poles, if anything. And these ions moving toward the Earth would not impede incoming cosmic radiation like ions moving up.

More evidence we have for more charge in the north is that the Earth's magnetosphere is imbalanced to the south. The magnetosphere is not the same size top and bottom, as it would be with a true dipole. I have seen this attributed to the tilt of the Earth and other factors, but obviously it can't be tilt, since the Earth is sometimes tilted toward the Sun and sometimes away. If the tilt were the cause, the shape of the magnetosphere would switch every six months. Again, imbalanced charge (parity violation of the entire field) is the most logical answer.

For now, we will return to the spin of the Earth. I said near the top that charge not only caused the spin of hurricanes and so on, it also caused the spin of the Earth itself. How does it do that? Simple mechanics, as usual. All my photons, including charge photons, have real mass and angular momentum. Even standard-model photons have real momentum, for if they didn't we wouldn't have a photoelectric effect. Well, during this recycling of the charge field, the photons have to be curved or redirected by the interior of the Earth. I said above that this was done by fields, but that was shorthand, of course. In my theory, fields like this are collision fields. The entire charge field is a collision field, when you get right down to it. Just as with Feynman's sum-overs, what you have with photons is a stupendous amount of field collisions, and you sum them to get your overall motion. Some photons will go right through the Earth without a collision. Some will crash head-on into an anti-photon, losing spin and energy and being "demagnetized". But the median or defining photon will appear to create a nice curve, going from south pole to just above the equator, say. The many collisions this photon encounters will sum into this curve, and the bulk of the photons will follow that curve, more or less. It would take a lot of math to show that curve, and this paper is already overly long, but I think that curve is fairly intuitive, once you understand the mechanics. If you have a spinning Earth and spinning photons, and a dipole configuration with opposite spins coming in at opposite poles, you are going to get curves. I hope you can see that without all the math.

And so, given that, you only need to add the fact that these collisions in the Earth's interior transfer momentum and angular momentum. When the photons collide with matter in the Earth, that matter feels a tiny push. If we sum all the collisions, the Earth feels a force. It feels a force in the direction of motion of the photons, it is that simple. We don't have to do any mathematical switcheroos. So, due to linear momentum, charge coming in at the poles tends to make the Earth a bit smaller, and charge going out at the equator tends to make the Earth a bit larger. That is the answer to that question, not the given one. The radius at the equator is greater due to photon pressure from within. And the Earth is flatter at the poles due to photon pressure from without. The angular momentum of the photons is transferred to the Earth in collision as well, causing spin.

This would create spin even if we only had photons coming in at one pole, but photons coming in both poles doubles the effect. Since I have shown we have more photons than antiphotons, we must have more flattening at one pole than the other. In fact, this is precisely what we find. The south pole has a fraction more flattening than the north pole, and this is the cause. The south pole is being flattened by the same cause that obliterates the nearside crust of the Moon: charge photon bombardment. Interestingly, I predicted this flattening before I knew of it. I wrote it into this paper, and only then Googled on it. Fortunately, I found this‡ and much more. If you prefer the current answer to greater radius at the equator, consult the current answer for an explanation of more flattening at the south pole. The Earth's angular momentum obviously can't answer that one, nor can centrifugal forces. And gravity from the Moon can't answer it either. Charge is the pretty obvious answer, to this as well as to many other questions.

If you take that last link, you will find that we also have a small bulge at the north pole. Can current theory tell you why? No. But I saw the answer immediately. Since the ambient or Solar field is not balanced in terms of charge (see also my paper on parity violations), this causes an imbalance in matter/anti-matter. Yes, the Earth has more matter than antimatter (though it does have antimatter). This means that anti-photons are coming in at the north pole, there meeting a body composed of matter. This is not disallowed, but it does create a local field response. The incoming charge cancels the local charge, as a matter of spin, and the flattening that would normally take place is damped down locally. The flattening effect of the incoming particles is lessened, since they don't have an angular component to their momentum. This makes the local surface seem to rise relative to the area around it.

*http://www-paoc.mit.edu/labweb/lab5/inertial
%20circles/inertial_circle.pdf
**http://en.wikipedia.org/wiki/Earth's_magnetic_field#Field_char
acteristics
†http://www.21stcenturysciencetech.com/Articles_2011/BigBang
_Bunk.pdf p.3
‡http://www.pd.astro.it/E-MOSTRA/NEW/A2012ERT.HTM
•http://www.hulu.com/watch/81732/3d-sun

Chapter 14

The
CHARGE FIELD
causes the ICE AGES

Since I am not a geophysicist, I had never really studied the math for the ice age cycle. I only stumbled across it when I was writing my latest paper on the Sun. At Wikipedia, I found this enticing tidbit on the "Sun" page:

A recent theory claims that there are magnetic instabilities in the core of the Sun that cause fluctuations with periods of either 41,000 or 100,000 years. These could provide a better explanation of the ice ages than the Milankovitch cycles.

Yes, they could, since, as I will show presently, the Milankovitch variables are garbage when it comes to explaining long-term cycles. However, any theory that proposes magnetic instabilities in the core of the Sun as a mechanism for ice ages would have to explain the *cause* of the magnetic instabilities. I have studied the "recent theory," and in that regard it is a ghost. The theory is not really a theory, it is a model, because it shows effects but not causes.

This is the theory of Robert Ehrlich, who had an article in the January 2007 *JASTP*.*   He bases his theory on nothing but a computer model, by which he shows that magnetic instabilities *could* cause small temperature fluctuations in the core of the Sun. I have nothing against his model (at this time), and it may well be correct in its outlines, but he offers us no good cause of the magnetic instabilities, which I think we can all agree is a big hole in the theory. Others have proposed quantum effects (of course!) to explain these instabilities, but they are worse than ghosts. They are, as usual, flights of fancy and bad math.   I will show with simple math and mechanics that the variations in the Sun are caused by outside influences, not by processes within the Sun itself.

The theory that Ehrlich is trying to improve upon is called the Milankovitch theory, a theory from around 1914. It is based on long-term cycles in the Earth's eccentricity, tilt, precession, and so on. Strangely, though we are told on the "Milankovitch cycle" page at Wiki that "the theory has overwhelming support," it is admitted on the "Ice Age" page that Milankovitch cycles "probably cannot start an ice age." This is because the longer term cycles, though larger, are said to be caused by smaller variations. For example, it is admitted that eccentricity has a smaller effect on so-called Solar forcing, and yet eccentricity is the variable that matches the 100,000 year number. Obviously the ice ages are the largest effect. How could the largest effect be caused by the smallest cause?

A bigger problem, not admitted at Wiki or anywhere else, is that variables like eccentricity cannot *cause* anything, since they are field effects not causes. In other words, there is and can be no "Solar forcing," since the Earth does not force the Sun to do anything. It is the other way round, of course. It is the Sun that does any forcing. It is variables transmitted by the Sun that cause all the terrestrial variables, including eccentricity, tilt, and so on. Smaller bodies do not force larger bodies; larger bodies force smaller ones. I will be reminded of the old equal-and-opposite rule, but if the Earth responds to the Sun in kind, that response will still be swallowed up and ignored. All effects, whether gravitational or E/M from a body the size of the Earth to a body the size of the Sun are negligible. They cannot cause these large effects seen in the ice core samples.

The Milankovitch cycle has many, many problems, most of which are admitted, but the greatest problem is that all the proposed effects together can't come near explaining what we see. The theorists then propose feedback mechanisms to increase the effect, but it is much more likely that the effects from these variations are actually damped by other variations (like greenhouse gases), rather than amplified. This is admitted even on the Wiki page, where one of the strongest causes of variation, axial tilt, is admitted to be resisted by other environmental variables, including greenhouse gasses.

It became apparent to me very quickly that this question was like all the other in physics and geophysics: the current answer was very poor, everyone seemed to recognize that in moments of candor, but the current answer was nonetheless guarded as a precious thing, since so many careers had been built on it. I could tell at a glance that the Milankovitch cycles were jerry-rigged and pushed, since they had no structural soundness from the first tap. Math is a lot like architecture. You don't have to study the Louvre for many weeks or years to see that it is more structurally sound than the Pompidou, for example. You can tell at a glance. It is the same with the math of these theories. Good theories are simple, and the math has no tape on it. Bad theories are full of paste-overs, pushes, and props, and they always come with a long list of assurances, insurances, and apologetics. They also come with complete their own cops, who will threaten and bully anyone who points out the tape and the props. Good math and theory doesn't need cops: it is its own recommendation. Only bad theory needs to intimidate you to believe it.

I was also amazed at how simple the right answer was, as usual. The right answer had been missed not because it was so complex and esoteric, but because the current physicists had preferred to bury their heads up their own black holes one more time. As I will show in just a moment, the answer depends only upon seeing influences from outside the Solar System, and we should know of those influences. We are not ignorant of the galactic core and its incredible power. So an impartial observer will ask why physicists are so blindered when it comes to admitting input from beyond the Solar System. The math and mechanics I will show you are hardly revolutionary. But physicists don't want to go there because this input must come in on the charge field, and present physicists are allergic to the charge field in celestial mechanics. They don't want to even consider it, because it will mess up all their work, all the way back to Laplace. Yes, they will have to re-do centuries worth of math, and they don't want to do that, the truth be damned.

But the mechanics is fairly simple. It takes about 240 million years for us to circle the galactic core. If we divide that by 100,000, the time between ice ages, we get 2,400. So the Earth does something 2,400 times in every galactic orbit, and that something causes either cooling or heating. What could it be? I suggest a Solar System alignment with the galactic core, which would align the galactic charge field with the Solar charge field. For the galactic core to augment the Solar field like this, the ecliptic either has to be in the same plane as the galactic plane, or the nodes have to be perpendicular to the galactic core. Since the ecliptic is now at a large angle [60°] to the celestial equator, those planes don't match. Instead, we must study the nodes.

Some will think this is astrology, due to the terminology, but it isn't. It is straight mechanics. The nodes I am talking about are just the two points on the circles where the ecliptic meets the galactic plane, as in this illustration.

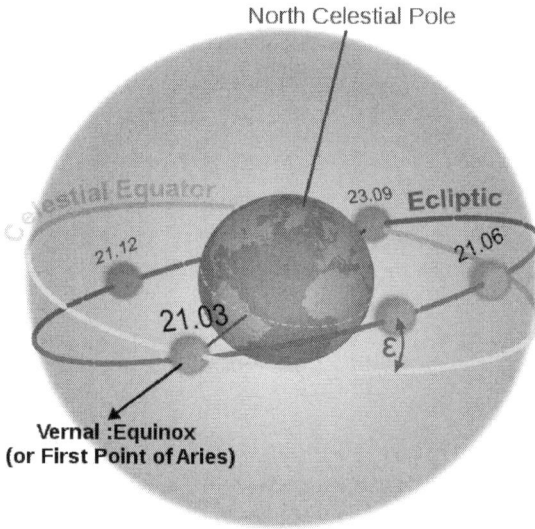

This illustration is from the Wiki page on the ecliptic, not from an astrology site, so stay calm, please. It is showing the plane of the Solar System relative to the Earth's equator, but we can pretend "celestial equator" means galactic plane if we like (since I couldn't find a diagram of that, this will do). We just change the angle from 23 to 60. To be even more precise, we would change the ecliptic plane to the equator of the Sun, since there is a 3° difference, but the angle doesn't actually matter in this problem so I will skip all that. All we need to know for now is that there is an angle and that the nodes travel.

It also doesn't matter if the Sun is actually in the main galactic plane or not. All we need for this theory is the actual plane between the galactic core and the Sun, since points outside the main plane receive charge just as do points in the main plane. They don't receive as much, but they receive plenty. We are told that the Sun travels from below the plane to above it over very long time scales, but that also doesn't matter here. It won't affect this paper. It may answer even longer timescale problems, but it doesn't affect this one.

As it turns out, the nodes move in the same sort of precession that the precession of the equinoxes does, since this precession causes that precession. That is why I can use this diagram. It is currently believed that so-called lunisolar precession is caused "by the gravitational forces of the Moon and Sun on Earth's equatorial bulge, causing Earth's axis to move with respect to inertial space," but that is false. Since I have shown elsewhere that gravity is a motion, not a force, precession cannot be caused that way. Einstein showed that gravity was not a force, and although current physicists accept that, they haven't let it sink in too far. If gravity is not a force, it cannot cause precession in this way. But this probably deserves another paper, since I haven't addressed it yet. You may simply notice for now that if I am correct, this motion of the Solar plane relative to the galactic would cause a precession of the equinoxes with no real motion of the Earth's tilt. To decide the question, we only have to study the **Solar** precession of other planets. If I am right, they should all precess on the same timescale. I found data on other sorts of precession for the other planets, but none on this. Either it isn't known or it isn't widely publicized.

Now, go back to the diagram. We will use their nodes, as a convenience. As you can see, one of the nodes I am talking about used to point at Aries, hence the name. It now points at Pisces. It travels through the zodiac, taking about 23,000 years to do so, we are told. When I say that the Solar System should be perpendicular with the galaxy, in order to cause a charge conjunction, I mean that the line running through the two nodes is perpendicular to the galactic core. Since the galactic core is in Sagittarius, this means the nodes would be pointing roughly at Pisces and Virgo (right angles to Sagittarius). Since the nodes *are* pointing at Pisces right now, we are in an interglacial period. In other words, I will show now that we are in an interglacial period *because* the nodes are pointing at Pisces.

It is actually very simple, and completely mechanical, but to explain it simply and visually is a bit tricky. I suggest you grab two CD's or DVD's or other small disks. Hold one in one hand and one in the other. Hold the one in your left hand horizontal, or flat to the ground. That will stand for the galaxy. The one in your right hand is now the Solar System. The Sun is at the center of the disk and the Earth is part of the way out, orbiting. Now, the Solar disk is not flat to the ground. It is at an angle. Over time, we will let this angle stay the same, but we will move the high point. Start out with the high point of the disk pointing toward the galaxy disk. When that is the case, the nodes are not pointing at the galaxy disk. You can see this if you bring the disks together. If you could superimpose them, as in the diagram above, the nodes would be pointing to the sides. Now let the high end of your Solar disk go ¼ turn either way. In that case, the nodes are now pointing at the galaxy disk. I will show that when the nodes are perpendicular the charge conjunction is at a maximum.

Over time, the high end of the Solar disk makes a full revolution, returning to its original position. This is one cycle, and it is this cycle that takes about 23,000 years. But the nodes will be in line with the galactic core in two positions: when the high end is at ¼ and ¾. Or, if the galactic core is north, the nodes will line up with it when the high end is east or west. And so we get an alignment every 11,500 years or so. Alarms should be going off in your head now, because that number is already an important one in the history of ice ages. According to the math of many, interglacials have lasted about 11 thousands years. This is where that number comes from. 11.5 is half of 23.

If that were all there were to it, then we would get a warming from this alignment every 11,000 years. Actually, we do. If you study the ice core chart from Vostok in the Antarctic, there are, yes, 9 little peaks in each of the longer periods, and the peaks are an average of about 11,000 years apart. All of those peaks indicate a warming period. We aren't usually told this. We are told the interglacial periods are 11,000 years long, but we aren't told that there are 8 other (sub)interglacial periods, all of them also about 11,000 years long.

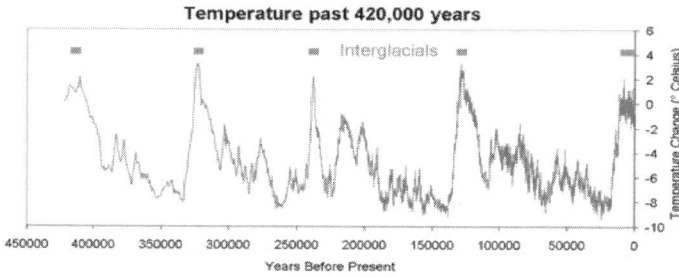

Temperature past 420,000 years

If you want to confirm this yourself, just take your ruler to the chart. You will find a very clear 3/16 inch gap between many of the adjacent minipeaks. This corresponds to about 11,000 years.

But we still have the longer period to explain. There must be another variation, because some of these warming periods are a lot larger than the others. Only one in nine of the warming periods is large enough to melt the glaciers and be called "interglacial." And this 1 in 9 comes (almost) like clockwork.

What causes it? Jupiter. Jupiter causes a wobble in this cycle, sort of like the wobble in the tilt of the Earth, due to nutation. Jupiter provides a nutation to the motions above. Because of Jupiter, one of the nine conjunctions is cleaner than the other eight, and this causes a greater maximum in the charge field.

Remember that even mainstream physicists have pointed out that our system is almost a two-star system. Jupiter is nearly as large as many red dwarf stars, and he recycles nearly as much charge as a small star. Since we are studying charge

213

here, not fusion, it doesn't matter that Jupiter is not fusing. Only the charge matters. Therefore a charge alignment must include Jupiter. The important fact here is that Jupiter is not precisely on the Solar plane or the ecliptic or the invariant plane or the Solar equator. Once again, we have a small angle. This angle causes the wobble. It causes it because it now matters where Jupiter is in his inclination cycle when the galactic alignment occurs. In other words, at each 11,000 year alignment, Jupiter is in a different place. When Jupiter is nearest the Solar plane, he most augments the charge maximum. When he is most off the plane, he damps it. But because of the way his inclination matches up with the larger cycle, he is only at his nearest point about every 100,000 years.

Some will say the line is the invariant plane, in which case Jupiter is nearly on it already. But the charge field is not determined by the invariant plane. It is determined by the Solar equator, through which most of the charge in the Solar system is cycled into the system. Therefore, the system is most efficient when Jupiter is crossing the Solar equator or is nearest to it. Jupiter is currently 6 degrees off the Solar equator, so charge efficiency in the system is not near the maximum in that regard (unless Jupiter is usually more than 6 degrees inclined).

Again, I couldn't find good data on the precession of Jupiter's inclination. It appears that the current number for the cycle is about 50,000 years. That is nice, because it is half our long cycle. Both an old book on Google books and a new video on youtube used that number, though neither mention gave much detail. In neither place could I find out whether the number came from data or from theoretical models. In any case, we could have predicted we would find the number 100,000 or a simple fraction of it. Since Jupiter must be the cause, we know that the numbers will work out one way or the other.

Some will find that last statement peculiar, but once again I solved this one because I knew where to look (and where not to look). I knew the Milankovitch cycles couldn't explain this, because bodies don't cause effects upon themselves. Just as the Earth cannot be "forcing" the Sun, it cannot be forcing itself. Local mechanisms can affect eachother, and I am not denying it, but large long-term cycles like this cannot be caused locally. Milankovitch was looking in the wrong place from the start. His opening postulates were illogical. He was trying to explain effects via other effects, and that can't work no matter what you are looking at.

For some reason, humans are not yet adept at looking beyond their own environs for mechanical explanations. Our sight is still generally very limited. We know that the Sun causes everything here, and we should know that the galaxy causes the Sun to do whatever he does, but we aren't good at peering up the line of influences. Some of us have prayed to the Sun, but I don't know of a people who have prayed to the galactic core. My solution to this problem tells us that would have been the logical thing to do. Supposing that powers greater than us required prayer or worship, we should have been worshiping the great deity seated in or near Sagittarius, who plugs in our Sun and thereby powers everything in this system.

But enough narrative color. Let us return to the straight mechanics. I have some large questions yet to answer. The first one is, "Supposing you are right, how, precisely, does the charge field of the galaxy align with the charge field of the Solar system? You have a sixty degree angle, no matter what. Why should one configuration give us maximum input, and warming, and another configuration give us minimum input, and cooling?" Well, we need to know how the Sun acts as a conduit of this energy, to understand how the fields hook up mechanically. Before I discovered how the Sun worked, it could have been argued against me that the charge from the core of the galaxy simply arrived like other light does, coming directly from source to receiver. In other words, we don't require that visible light be cycled through the Sun. The light from Sagittarius goes directly from there to here, with no stopover in the Sun. But it turns out charge doesn't work that way. Charge is made up of lower energy photons, and these lower energy, longer wavelength photons do make a stopover in the Sun. They get sucked in by the charge vortex created by the Sun's spin and his charge potentials. A large number of them go in the Solar poles and are recycled. I don't know yet if the Sun borrows some of their energy, spitting them out at longer wavelengths than were emitted, although that is a good hypothesis. What I do know is that the Sun emits them more heavily at the equator, due only to angular momentum. The greatest velocities are at the equator. Therefore the charge is heaviest in the plane of the Solar equator, and that applies all the way out to Pluto. The planets live in or near this plane because that is where the energy is. The charge potentials push them there, and by being there they are constantly energized by the field.

Still, that explains very little, at first glance. We have that 60° angle to explain, and we cannot get rid of it no matter how we look at the input. No, we can't get rid of it, but we can discover how its influence varies. We can study our disks again, to see how the variation works. At first I thought the maximum conjunction would happen when the nodes pointed at the galactic core. When the nodes are pointing at the core, the angle isn't important. This is because the angle is sideways to the influence, and so it isn't really an angle in the field. To see this, just position your disks. If the galactic disk is N, let the high end of your Solar disk be east or west. In that case, you have an angle, but the galactic core doesn't see it. The two disks are tilted relative to one another, yes, but the angle is 90° to the field. When one field influences the other, the angle is "invisible." It won't affect the mechanics. If you don't like the word field, we can look at the photons. When the photons arrive from the galactic core to the Sun, they don't experience that angle. They don't have to travel through it. They don't care whether the Sun is tilted or not, since he is not tilted at them. He is tilted to the west, say, and they are coming from the north. Straight mechanics.

This is why I thought this configuration would cause maximum effect. The angle is put out of "sight." Normally, I would have been right. This configuration would have caused maximum conjunction of the fields. In the opposite configuration, the high end of the Solar disk is N or S, and the angle is pointing either directly at or away from the galactic core. So the core "sees" the angle. The charge coming from the galaxy hits this angle, and we have to take a sine of that angle to calculate the effect. That would normally produce a minimum.

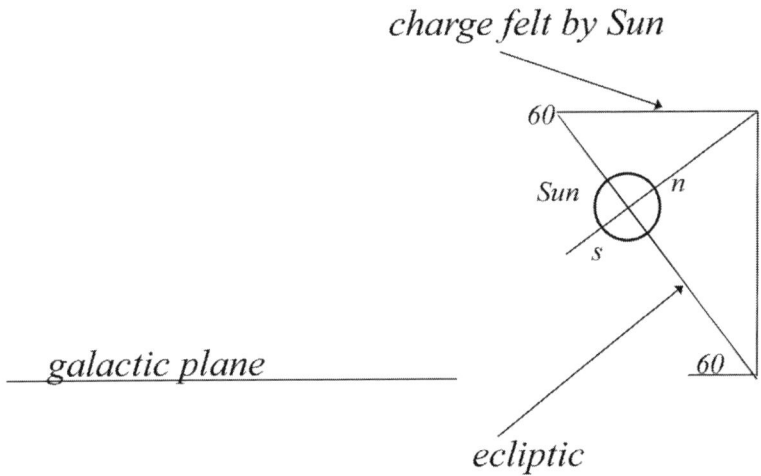

charge felt by Sun

60

Sun n

s

galactic plane

60

ecliptic

But I had forgotten one very important thing. The plane of the Sun here, that we are representing as the disk, is the plane of the Sun's *equator*. But the Sun doesn't receive charge at the equator, he emits it there. Which means we need to align the galactic charge field with the Sun's *poles*, not the Sun's equator. This reverses everything we just assumed about the disks and the maximums. When the charge from the galaxy is coming in at the equator, the field conjunction is at a minimum. The emission of the Sun counters it, or damps it, and much less of its makes it to the poles. But when the angle is in line with the galactic core, we have a maximum. We have to take the cosine of the angle, to figure out how much of the galactic field actually pours into the poles, but that portion of the field is the most the Sun can possibly get. The Sun never tilts 90° to the galaxy, so he cannot receive the full dose. Cos60° is the best he can do.

To say it another way, let us look at the photons. If the photons from the galactic core arrive at the Solar equator, they are met by a heavy barrage of photons coming right at them. We get a lot of spin cancellations and potential cancellations, and the two fields of photons don't stack. Mathematically, they cancel. But if the Sun is tilted toward the incoming photons, they can more easily pour right into his pole. Not only do they experience less field traffic from emitted charge photons coming at them (since the Sun emits less nearer the poles), they also find a straighter path into the pole. It is almost like the Sun leans over to let them in the hole, you see. And this works either with the north pole or the south pole. In one case it is photons pouring in and in the other it is anti-photons, but neither the Sun nor the galaxy really cares which it is. The Earth doesn't care either, since we don't even know the difference here.

As you now know, this matches data, since we are currently in this second configuration. The angle *is* in line with the galactic core, since we know that the nodes are pointing at Pisces. And we know we are at or near a maximum, because we are interglacial. Jupiter must also be nearer his maximum, because we are still quite warm. We are at zero instead of 4 on the chart, but we are way above -9. We take this to mean that Jupiter being inclined to 6° is nearer his maximum than his minimum.

Next we have the mechanics of the Jupiter libration. I have done the overview but haven't explained to you the specifics. How does the charge field of the Sun meet that of Jupiter? Well, since Jupiter's poles are right side up, like the Earth but not like Venus, we assume he has the same charge profile as the Sun. Meaning, he is mainly matter not antimatter, and photons not antiphotons. If so, then he would amplify the charge from the Sun, both the magnetic component and the electrical component. The closer he is to the Sun's equator, the more he amplifies, since we have simple poolball mechanics here. I would say that is the best assumption at this time, but the reverse situation would also fit the theory. If we discover from data that the charge of Jupiter actually damps the charge of the Sun, then our maximum charge conjunction is when Jupiter is furthest from the line, rather than closest to it. Either situation creates maxima and minima, on the same timescale, so either would work.

As for the timescale, I can't address the analyze the data without knowing what it is. But let us assume for now that the inclination cycle is currently around 50,000 years. In that case we have to explain the extra 2 in the math. I suspect that the number 50,000 is for the move from maximum to minimum, not the full cycle, in which case our math is done. But there may be other easy explanations of the number 2, if that should turn out not to be true. For instance, since the Earth reverses polarity occasionally, Jupiter probably does too. These pole reversals might add a 2 to the field, since we would then have two possibilities over the long term: N and S. This would automatically double the long-cycle number. There are other possibilities, but I will not complicate this paper with more analysis. When I have firm data, I will do more work on this.

We are told that we are late for the next ice age, and if you study the ice core chart you quickly see where that conclusion comes from. However, if you study it with a bit more rigor, you will see that we have been late for the next ice age for more than 20,000 years. If we are just going on the ice core readings, and a statistical analysis, this current warm spell should have started around 40,000BC, and should have ended around 25,000BC. This means that Jupiter is not doing his old job like he used to. I would say that it is very likely that Jupiter is not exactly what he was 400,000 years ago, or even 100,000 years ago. In other words, it is very likely that the inclination cycle of Jupiter has changed in the past half million years, due to taking on new moons or to other Solar System changes. Since Jupiter is the cause of the long cycle, a small change in Jupiter's inclination would change the whole chart dramatically.

We also know from other charts that Jupiter cannot have been what he is now in the time of the dinosaurs. The dinosaurs existed for around 160 million years, so this ice core chart obviously doesn't go back that far. What I mean is, there couldn't have been ice ages every 100,000 years back then, so the conjunctions we have now are different than the conjunctions we had back then. Either Jupiter was more stable, with no or very little inclination variation, or the Sun was tilted more with regard to the galaxy, or something. As I will show in another paper, most likely the Sun was nearer the galactic plane proper, and so it received more charge all the time. Whatever the cause, we see that things change. Charts change, and charts change because bodies move. This ice core chart is telling us that things have changed in the past half million years. What happened before didn't happen this time, and our job is to find out why. Are the periods getting longer because Jupiter's inclination cycle is slowing, or are we entering a period of no ice, like the Triassic, because the Sun is increasing its tilt? Or is the Sun moving closer to the plane of the galaxy, or further away? These are the real causes of long-term temperature here on Earth, not global warming or anything else humans may do. We can only react in small ways to these changes. Burning all the fuel in the world can only delay a temperature change for a few years. If the ice age is coming, we may have delayed it for a couple of decades, but we cannot have delayed it for 20,000 years. According to the ice core statistics, it should have happened thousands of years ago, and we weren't doing anything then to delay it. From that alone we should infer our own unimportance.

We should cut our pollution so that we can breathe better, eat better, and live with fewer illnesses. And we should control our populations so that we don't all have to live in squalor (and so that other animals can live as well). But temperature is ultimately beyond our control. Unless we can move the Sun or Jupiter, we are out of luck.

That said, we had better begin studying more closely the inclination of Jupiter and the relation of the Sun to the galaxy. We can't continue to live in ignorance of the charge field, or the way the galaxy plugs into the Solar system. We can't respond to coming changes if we don't know what they are. I would say that currently we haven't got a clue. Since we are still explaining the precession of the equinoxes by gravitational magic, we must still be in the dark ages, mechanically. Since we haven't recognized the charge field, we must be living in the dark ages, mechanically. Since we haven't recognized the galactic input, which drives everything, we are in the dark ages, mechanically. It is way past time we quit larking around with black holes and the first seconds of the universe, which are problems way beyond us, and start looking at these problems that will affect us greatly, especially if an ice age is going to start any minute. We are told by various groups that we are either on the edge of a new ice age, on the edge of Armageddon, or on the edge of a flood caused by warming, but none of these groups are doing anything about it (except proposing new taxes and building new jails). We hear a lot of urgency in voices, to create alarm, but we see no action. If things were as dire as we are told, we would expect people to be rushing into action, as in the movie *Deep Impact* or something. Instead, we see physicists blowing billions of tax dollars jacking around with colliders, looking for hypothetical Bosons to fill holes in their mattresses, I mean matrices. And these same physicists can't even tell you what charge is, much less locate it in the field equations.

Once again, I have shown you that mechanics is the answer. I haven't discovered the whole answer yet, and don't expect to. But I have shown you the framework for the right answer. Giving charge a real presence in the field, giving the photon a real presence in the field, and finding the charge field in the field equations of Newton and Einstein are the pillars of this framework. The fourth pillar is keeping that charge mechanical, by explaining every motion and every force and every interaction in terms of collisions—collisions that can be diagrammed. No borrowing from the vacuum, no broken symmetries, no virtual particles, no undefined fields, no forces at a distance, no hidden variables, no hiding behind the math.

*http://arxiv.org/abs/astro-ph/0701117

chapter 15

The Hole at the Center of the Sun

Despite the intentionally provocative title, this chapter is not about an actual hole at the center of the Sun. It is about a hole at the center of Solar theory. If you read all the current material on the formation of a star, you soon realize there is a problem. We are told that stars exist by fusion, turning hydrogen to helium (in most cases) to create their energy. Fair enough. Since fusion requires high pressures and temperatures, we are told that stars exhibit such pressures and temperatures. That is easy to believe, since the Sun looks plenty hot. Problem is, we are told the current heat is generated by fusion, and we need the heat *before* the fusion. We are told that gravitational collapse creates high pressures, which created the necessary temperatures, and that is also easy to believe. However, we need a mechanism for that creation of high temperature from collapse, and as it is, we don't really have one. We can see that large celestial bodies become stars and small ones don't, so we are told that gravity starts the process. The star collapses and this creates heat and the heat creates the fusion.

But without more theory, that doesn't really fly. Why? Because the Sun isn't dense. Stars aren't generally dense. You would expect something that had collapsed to be very dense. But the Sun's average density is ¼ that of the Earth's. We are told that the Sun's core (where fusion takes place) has a density 150 times that of water, but even Wikipedia admits that is just a model. It turns out that it is a curious model, because in order to give the core that much density, the model gives the other 4/5's of the Sun a density of almost nothing. Even the lower photosphere, the level just above the core, is given a density of only $2 \times 10^{-4} kg/m^3$. That's 6,000 times less dense than the Earth's atmosphere. I would say that is grossly counter-intuitive.

It is illogical as well, since a gravitational collapse could hardly work that way. Let us say that the core is the real body, and the other 4/5's of the Sun is like an atmosphere. That is how the current Sun is sold to us, in the literature. Even then, the gravity of the core would have to act on the atmosphere more strongly than that, creating more atmospheric pressure and therefore more density. The core of the Sun is much larger and denser than the Earth, and yet the Earth's gravity creates a density in its own atmosphere of about $1.2 kg/m^3$, some 10,000 times more density than the Sun's photosphere as a whole.

I will be told that the energy of the fusion percolating up counteracts the gravity of the core acting on the rest of the Sun, but if mainstream physicists propose that, they are already admitting a unified field. I hardly think they want to do that, because that would be admitting I am right. If they are going to do that, I can stop this paper now. What I mean is, they *can't* propose that, because they have no mechanism for it. According to current theory, you can't turn off or cancel gravity, not by an E/M field, and especially not by an ion field. Gravity is gravity, and it doesn't matter how many ions or photons are flying up through the field. Niether Einstein nor Newton gives them any mechanism to cancel gravity, and QM and QED don't either. Therefore, this low density photosphere they are proposing is unsupported by current fields, to say the least.

Another problem is that a density of 150 times that of water is still not very high. Platinum at room temperature is 1/7th that dense. We can actually heat and pressurize platinum to densities nearing that, so it isn't that extraordinary.

All the theorists have really done is rig the density numbers to support their theories. They think to themselves, "What is the lowest density most people will buy for the rest of the Sun? We can't get it too low or people might start asking questions." Then they give the rest of the density to the core, and hope that density will impress you. But the fact is, an average density of 1.4 for the Sun isn't impressive no matter how you slice it. Without a whole lot more theory, you have to be really negligent to accept that that sort of density can start fusion, by itself.

Now, the electrical Sun people will say, "Right, so get rid of fusion altogether!" But I happen to think we have some pretty strong evidence for fusion. Neither the mainstream fusion theorists nor the electrical Sun people can explain all the phenomena we see, so I would like to combine them both. If we have both, we can explain more of the data. So I would like to propose that they are both right. All of my work on the charge field tells me that we do have fusion (see below for immediate proof), but we need charge and E/M effects to get it started. In other words, a star isn't born in a gravitational collapse, it is born in a unified field "collapse," where the charge field undergoes changes like the rest of the field.

This helps greatly in the current problem, because I have shown that the charge field has a mass 19 times that of baryonic matter. Every proton is recycling a photon field that outweighs it by 19 times. So when you add the charge field to any problem, you get a greatly multiplied effect. When you add pressure to any mass, you also add pressure to the charge field. When you squeeze the protons in hydrogen, you aren't just squeezing the protons, you are squeezing the billions of photons.

We should have known this decades ago, since we have known for quite a while that the Sun is a plasma. A plasma is an electromagnetic entity. It cannot be explained with gravity. To deflect this obvious reading of the facts, we are told that the plasma is created by the fusion process, but we have plasmas between here and the Moon, caused by the local charge and E/M fields. It doesn't require fusion to create plasmas, it only requires lots of ions in a charge field. It is not fusion that created the plasma in the Sun, it is the dense plasma that created the possibility of fusion. Once again, current theory is upside down.

In short, because the Sun was NOT dense, but contained a large amount of matter, it coalesced into a very large sphere that was able to recycle very large amounts of charge. You don't want too much density in a star, because the radius is more important than the density. The bigger radius gives you more angular momentum, which allows you to recycle more charge, which allows for a hotter plasma. All this charge passing through the sphere created a hot plasma, and the hot plasma allowed for the beginning of fusion. In this way, we see that much of the heat of the Sun predated fusion. And in this way, we see that the electrical Sun people are right. A good portion of the current heat of the Sun is *still* caused by the charge passing through the Sun. Fusion only adds to this heat. We can now (with my theory) even calculate the percentages of heat that come from fusion and from charge —see below for the math.

I will be asked why I am allowed to propose this when the mainstream theorists aren't. Two reasons: 1) I have the fields to justify it, since I have unified gravity and charge. 2) Even though I have a unified field, and can show how charge works in the opposite direction to gravity, I am still not using my unified field here to cancel gravity. I think current theory has the densities in the Sun all wrong, so I don't think we need to explain why the photosphere is so tenuous. There is no reason the core has to have all the density, so we can give more density to the rest of the Sun. Therefore there is no density split to explain. Fusion *doesn't* cancel the gravitational effects of the core upon the photosphere, so I don't have to explain *how* it does.

Concerning 1), I can propose the Sun as a unified field beast, because I have a unified field. Mainstream physics doesn't have a unified field, so they can't borrow any of this from me, in any part, without being grossly inconsistent. They have told us for 300 years that the celestial field, including the Solar field, was a gravity field only. So they cannot start slipping in E/M field assumptions here to suit themselves. If they want to browbeat anyone who so much as whispers "ether", fine, but after they have done that for decades, they cannot prance in later and begin telling us E/M effects in the Sun affect the gravity field of the Sun. If E/M or quantum effects cause changes in the gravity field, they have to show a mechanism and a field math.

[This also applies to black holes, of course, which are supposed to be collapsed superstars. The big theorists like Hawking and Penrose propose quantum effects in black holes almost daily, and quantum effects are E/M effects. Why does nobody ever ask them how E/M effects enter the field equations, which are gravitational equations, not quantum or E/M equations? When someone like me proposes that the field includes E/M, I am shouted down with a chorus of derision, told that there is no room in the field equations for any corrections. When someone like me proposes that Relativity is wrong by 4%, I am shouted down with a chorus of derision, told that there is no room in the field equations for any corrections. They are already correct to within a billionth of a nanometer or something, I am assured. But then Hawking and Penrose and all the other big names propose quantum effects in black holes, and no one bothers to tell *them* that there is no room in the field equations for E/M. **Quantum effects are E/M effects**, and if there is no E/M in the celestial field equations, Hawking and Penrose can't propose quantum effects to fill mathematical holes!]

I can propose E/M and quantum effects in the Sun all day, if I like, since I have shown exactly where the charge field fits into Newton's field equation (and therefore into Einstein's field equations). When I propose the one field affecting the other, as when the E/M field affects the gravity field, I have equations and theory that show how it is done. I have a coherent math, a coherent field, and a coherent theory. All current theorists have is wild assertions, based on wishful thinking and a near-infinite disregard for the intelligence of their audience. They feel free to propose quantum effects to fill all the gaps in their models, even though they known darn well that they don't have a unified field. If they have no room in their "perfect" field equations for corrections, then where do they fit in these quantum effects? The readers of these gentlemen appear to believe that quantum effects are so small they don't have to be fit into the field equations, but that is false. Quantum effects that were so small they didn't affect the field equations would be too small to affect the bodies in the field. Current theorists want to propose quantum effects that affect real bodies but that don't affect the math or the field. We must suppose it is another example of virtual forces, whereby ghosts in the field can cause real motions.

This reminds me of a conversation I had with a reader recently. He pointed out that engineers love my papers, but for physicists they are "unmentionable." I had to laugh, since I am about the furthest thing from an engineer imaginable. I am not a practical person, I don't have a lab, I do very few experiments, and I am not too fond of machines. I simply like solving problems. I am definitely a theorist. But this separation of engineers and physicists by my reader made me pause. Engineers like my papers because they are mechanical. Physicists dislike them for the same reason. Odd, don't you think, that physics has divorced itself from mechanics, when they used to be synonymous. It is somewhat like the 19th century separation of doctors and surgeons. Doctors had much more prestige then, because surgeons used their hands. Surgeons were almost blue-collar! In the same way, anyone who concerns himself with mechanics is now seen as a lowbrow. A mechanic is little better than a grease monkey. This is because math has taken over physics, and the math has separated from the mechanics. The mathematician keeps his hands clean of physical problems (see my comments on Pauli in my paper "The Einstein-Bohr Letters"). He considers himself more pure and elevated. But like the 19th century doctor, he is in the grip of an illusion, an illusion fed by careerism and ego. Separating the math from the mechanics was not a move toward purity, it was a move toward magic. Once the equations have become divorced from the motions, anything is possible. That is not purity, it is mischief. We see this all the time now in the new math of new physics. Everything is now explained in terms of a virtual field or some symmetry breaking. But both ideas are more magic than anything else. They are neither mechanical nor physical.

If physicists want to be free to propose anything they like, they should stop calling themselves physicists and quit calling the field physics. After all, the word "physics" should have a sour taste for these gentlemen and women who have climbed out above all physical limitations. For them, the physical is just that residue of the math left after decoherence, a nasty by-product like a sloughed off snake's skin. One wing of science (and pre-science) has always wanted to cast off the physical, and that wing has long been in control. It would never admit it, but science actually shares this hatred of the physical with many of the religious people it debates. A prominent form of Christianity has always had a distaste for the physical, and all to do with the body. The same can be said for Islam, Judaism, Zen, Buddhism, and Hinduism. In the same way and for the very same reasons, new physicists desire to leave the physical behind them. Religious people want to climb out into some realm of pure consciousness, and mathematicians do, too.

The religious people flee the filth associated with the body, and the mathematicians flee the filth associated with mechanics. Mechanics is the math of real bodies, and current mathematicians have an instinctive revulsion in the presence of real bodies, be they human or celestial. They can't even stand to diagram them or visualize them, telling you that such visualizations only get in the way of the math.

It is this sort of illusory or delusional math that allows these people to propose a lot of contradictory things without noticing how contradictory they are. When your terms are all just floating in your head, you can't see the physical contradictions. When physics is based on computer models, no one is there to spot contradictions. Computers cannot spot theoretical contradictions. There is no program that alerts you when you have just grossly contradicted yourself (although there should be—it would make a mint and would always be beeping). Only when you see physics as real *physical* bodies, bashing into one another with cruel and filthy forces, will you see the contradictions. It is not we "engineers" who should be forced to do math without asking mechanical questions, it is these faux-physicists who should be forced to draw pictures, and to diagram them. They should be strapped to their chairs and forced for months to do nothing but label kinematic diagrams with variables, and to write simple equations for those variables. They should be strapped to giant orreries, where they can see and feel for themselves what forces and collisions really are. They should be thrown back into the mudpits of grade school with bags of marbles, where they can be reminded what the world really is. It isn't a computer model. It isn't a virtual field of wishes and manufactured symmetries. It is a hard and fast realspace of mechanics, where the bodies around you won't put up with fake equations.

Now, let us return to the problem at hand. Wikipedia gives us a few "present anomalies" of the Sun, which include the current dimming of the Sun, the loss of ½ the magnetic field, the loss of 3% of the Solar Wind, and the fall in sunspot activity. Of course the Wiki-police don't like these anomalies and are trying to take them down. They tell us they are "outdated," as if last year's data can immediately be jettisoned as no longer valid. They want to take them down because none of these can be explained by gravity or fusion. The Sun has not changed its size or mass, the Solar System has not lost or gained any mass, and so on. The only way to explain all these linked phenomena is with the charge field. The Sun is currently recycling less charge than usual because it is *receiving* less charge from outside the Solar System. Remember, our entire system is traveling through the galaxy at high speed (250km/s), in an outer arm. Well, the galaxy is not homogeneous: it has areas of higher charge and lower charge. These fluctuations cause fluctuations in the Sun.

Put simply, our system is not a closed system. We know it is not receiving great inputs of normal mass from anywhere, so it must be receiving fluctuations in charge. But this of course implies that the Sun is not running on fusion alone. It is running on charge. When the charge in drops, all the outputs of the Sun drop.

These "present anomalies" could not be more clear in what they are telling us. The fluctuations are huge, way too large to be caused by "quantum effects" or other jerry-rigged explanations. The Sun is fed from the ambient charge field, which is what the electrical universe people have been telling us for years. And since that is so, the current (interpretations of the) field equations cannot be correct. We simply HAVE to include charge in the field equations. The only way to include these huge corrections in our "successful" field equations is to do it as I have done it. Since I have shown that the charge field already exists in the current field equations, they don't have to be completely rewritten. They only have to be re-expanded and re-interpreted, to show which part of the old field is charge and which part is mass. Or, to say it another way, which part is baryonic and which part is photonic.

Although we cannot explain the current anomalies with fusion alone, we cannot explain them with charge alone either. The anomalies are proof of fusion as well. Why? Because if the Sun were based on E/M or charge alone, these passes through lesser charge would be catastrophic for us. Let us say the Sun is passing through a charge field that is much less than normal, as it apparently is. All the other outputs would have to drop by large amounts as well. The Sun couldn't lose all that charge and keep most of its light and heat. The Sun has lost only .02% of its light, according to the researchers, and 13% of its temperature. But if the Sun were electrical only, a big drop in charge would cause the immediate death of us all. We would immediately freeze. If the heat followed the magnetism, for instance, we would have lost half our heat.

To have kept its heat, the Sun must be storing energy. How is it doing that? Well, there are various answers to that, and the question is far from being decided, but an easy answer is that the Sun doesn't have to store energy to make it through these down times. Once fusion has started, it won't stop unless the temperatures in the Sun drop below a certain level. So fusion continues, even when charge inputs drop considerably.

From the numbers above, we can now calculate how much of the Sun's energy comes from fusion and how much from charge. If we take the numbers from Wikipedia as correct, we find that "Its magnetic field is at less than half strength compared to the minimum of 22 years ago." Well, that doesn't make any sense. You can't compare one minimum to another. They must mean it is at half strength compared to some maximum. But it doesn't matter, since the magnetism won't tell us anything here anyway. The charge field we are passing through may be less magnetic than normal, but still have the same charge density. Magnetism just tells us how the photons are spinning, not how many of them there are. So we would be better looking at other numbers. The temperature would appear to be an important number, but any analysis shows that temperature isn't a good indication of total energy output or of charge either. Temperature must be a function of both charge and fusion, so it won't help us isolate either one. And it won't follow total energy, since fusion will likely absorb more of the total temperature as the charge diminishes.

Density is probably the primary indicator here, since density fluctuation would most likely be a straight function of charge pressure. Since charge is 95% of the field, density should follow charge to within 5%. Wiki tells us the density of the Sun has dropped 20% in the last two decades, so the charge field has dropped about 20% in that time.

Of course those are just rough numbers, to show you the math. There is no indication that the Sun was at an all time high 20 years ago, and we need to calculate against a strong maximum to get a good number, of course.

What would be the best indicator of total energy output? Of the numbers at Wiki, I would choose the speed of the Solar Wind. The Solar Wind must be driven by both the output of fusion as well as by the recycled charge, so it is a fair indicator of the total field. The Wind has dropped by 3%, we are told, and if that is so, we can can calculate what percentages of energy output are caused by fusion and charge. If a 20% drop in charge causes a 3% drop in total output, then by this equation

$$(1 - x) + .8x = .97$$

we can find the fraction that goes to charge, which is 15%. That leaves 85% of the energy of the Sun produced by fusion. That makes sense, because it explains why all this loss of charge energy doesn't cause the Earth to freeze over like Neptune. The Solar System would have to pass through a very large pocket of low charge to affect fusion, since it would take the Sun quite a while to cool. In this way, fusion is the battery that stores energy: it is heat that takes time to dissipate. Charge isn't stored, but heat is. According to the equations we just ran, the Sun requires only 15% of its total energy to initiate fusion. The Sun has cooled by 13%, but it would have to cool by about 85% for fusion to cease. And since fusion creates 85% of the current energy anyway, the Sun wouldn't stop fusing even if we travelled through a large patch of zero charge. The charge would have to be turned off for a long time for the Sun to cool below 15% of its current energy.

By the way, this also explains charge reversals, including the magnetic reversal we are currently experiencing. These pockets of charge are made up of both photons and anti-photons, in varying amounts. Anti-photons are just spinning the opposite direction of photons, and they cancel the spin of photons in a charge field, canceling the magnetic field. If the area of charge we are traveling through had equal amounts of each, the area would be non-magnetic, like Venus. But, like Venus, it would still have the same amount of charge and the same amount of electricity. This explains the big variances in the magnetic field of the Sun, without equal variances in other variables. We are passing through areas with more or less anti-photons. But currently we are not only passing through an area with more anti-photons, as a fraction of the whole. No, we are passing through an area with more anti-photons than photons. If we were passing through a photon field before, we are passing through an anti-photon field now. In other words, the average spin on the photons has changed, changing not only the strength of the field lines, but the direction of the field lines. The poles of the Earth, which are just indicators of these field lines, must move with the field lines.

To say it a third way, the E/M field in and around the Earth is caused by the Earth taking in these photons or anti-photons at the poles and emitting them more heavily near the equator. As the photons go, the ions go, pushed along like a stream. But if the intake of photons suddenly switches to an intake of anti-photons, the spins in and the spins out interfere. Many spins are canceled, and the magnetism drops for a while. On the Wiki page for "geomagnetic reversal", we find this,

These events often involve an extended decline in field strength followed by a rapid recovery after the new orientation has been established.

Current theory can't tell you why that is, but I just did. Eventually the old photons will be used up and the Earth's field will match the ambient field. The field has gone from photonic to anti-photonic. Since photons and anti-photons are just upside down relative to one another, the field must go upside down, too. The Earth does not flip, only the field flips, and all this means is that the spins on the photons have reversed. Since anti-photons are not dark or evil or anything else, this spin and magnetic reversal is not any sort of catastrophe. Scientists admit that the Earth has gone through pole reversals many times without extinctions or inundations or anything else. It is predicted that the speed of the reversal may entail some big physical changes, but I have nothing to say about that. I am not a geophysicist. All I know is that anti-photons, like other anti-matter, are nothing to be afraid of. We live with them all the time. Anti-photons are all around you, and always have been. They are not cancerous, radioactive, or anything else. Anti-matter is just another form of matter. The pole reversal has already begun, and you have felt nothing. That is probably what you can expect to continue to feel.

Conclusion: Since the celestial field equations have contained charge from the beginning (since Newton, anyway), we are free to use charge to explain phenomena. And since the galactic core must be supplying the Solar System with large amounts of charge, we are free to use that fact to explain phenomena. Once we admit that the Solar System is not a closed system, and not a gravity-only system, most of the old problems evaporate. We then have a mechanism to solve centuries' worth of intractable questions. This is just one of them.

The Sun is not only a fusion reactor. It is both a fusion reactor and a charge synapse. Although a majority of its energy currently comes from fusion, a large percentage (about 15%) still comes from the recycling of charge coming from the galactic core. It also required the charge input to initiate the fusion sequence billions of years ago. It is this dual role of the Sun that explains all the "present anomalies" listed at Wikipedia, and hundreds of others not listed there.

Chapter 16

The

GOCE SATELLITE

provides more proof of my
UNIFIED FIELD

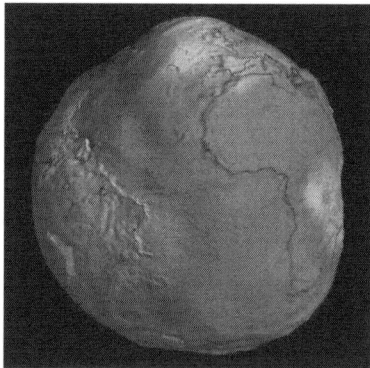

In news released on March 31, 2011 at the BBC*, we get more proof of my Unified Field. Above is the diagram they published from the GOCE mission in Europe. This is a schematic or model of the gravity field of the Earth, exaggerated to show variations. Yellow is most gravity and blue is least. You can't see the south magnetic pole in this particular view, but it is yellow like the north pole.

With a gravity-only theory, this is impossible to explain. Physically, we have flattening at the poles—the opposite to the above schematic—which means if you apply Newton's equations or Einstein's field equations to the problem, you get less gravity at the poles. Not only is there a greater radius at the equator—and gravity is directly proportional to radius in Newton's equations—but there is more mass at the equator as well. Gravity is directly proportional to mass, too. To explain data like this requires the mainstream to either torture equations, or to propose the unprovable: that the interior of the Earth is denser at the poles.

You will say, "What do you mean gravity is directly proportional to R? A greater R implies *less* gravity, by this equation: $g=GMm/R^2$. That is inversely proportional, not directly proportional!" Ah, but that is the wrong equation for this problem. We are not just moving an object to a greater radius, we are asking how radius affects the gravity field of an entire planet. And bigger planets have greater gravity fields, as we know. So the question is, if we made the Earth larger, and let it keep its current density, would g go up or down? Everybody should know it would go up, according to current equations and theory. That is the math we need, not $g=GMm/R^2$. This means that g at the equator should be greater than g at the poles, according to current theory. There is more mass in the line of the radius under a point on the equator, you see, and that is what matters to Newton and Einstein. The only way this would not be true is if the density was *less* at the equator, but I will show that is not the case.

[I have had readers write in and complain, saying that I have it wrong. They insist $g=GMm/R^2$ is the right equation. Well, yes, if you use it right, it can be made to work; but it usually isn't used right. Here is what I commonly see happen: they say, "Since a point on the equator is at a greater radius, if we use that equation we find g is less." I say, "But that only works if you keep M the same." They say, "It is the same Earth in both cases, so how could M change?" I say, "You have to take into account the mass beneath the 'point' you are considering. It can't be the same in both instances." They say, "What?" I say, "Start with the point on the pole. Say it is at R=6360km. Then run the equation. Then go to the point on the equator. Say it is at R=6375km. If you run the equation with the same value for M, you are implying that the point on the equator is resting on 15km of air, or of vacuum. If you are assuming the Earth is a consistent object, without huge density variations (which you are, in running the equations like that), you must take into account the extra mass beneath the equator. Because the point on the equator is at a greater R, it must have more mass beneath it, any way you look at it. And this is true whether you look at mass along the radial line beneath the point, or if you look at total mass beneath the point (the entire Earth). By definition, the point on the pole can't feel the gravity of as much Earth as the point on the equator. This is due simply to the definition of radius. The radius in that equation is *defined* as the radius *below* the object, or between the point and the center. Since the number you use is 6360, the point on the pole can't feel the mass of the other 15km. If you change the radius, you must change the mass. If you don't, you are implying that your equator is floating 15km up in the atmosphere, in some geosynchronous orbit. Given that, you must see that M will increase faster than R.

Say we double R. By the inverse square we quadruple the effect, or decrease gravity by four times. But mass will increase with volume, by $4\pi R^3/3$. So if we double the radius, the mass increases by more than 8 times. Mass increases more than twice as fast as radius, which proves what I said above. *In this problem*, gravity is directly proportional to radius. If you increase the radius, you increase the mass, which then increases the gravity. In Newton's equation, gravity is inversely proportional to radius only if R is empty space. R here is not distance of separation, it is actually radius. I hit this problem with even more rigor in http://milesmathis.com/weight.html.]

Since a first running of Newton's or Einstein's equations would imply less gravity at the poles, what scientists have done is propose that the surface of the Earth at the poles is pulled on more strongly by gravity, actually pulling it down; and that the surface at the equator is pulled on less. That would indeed explain what we see, except for one thing: it is not supported by the field equations. We would need more mass at the poles to start such a motion, and we have no indication of that. In fact, as I said, less radius would imply less mass, given roughly the same density. That is, from what we *do* know, it is probable the poles weigh less than equal areas on the equator.

If we look at the magnetic north pole to start with, we know the crust is thinner there, since the pole is most often over water. There is not much land anywhere near the north pole, so, at least as far as the crust goes, we have less mass. But even stronger indications come from the equator. It has long been argued by the mainstream that there was more mass at the equator, as well as a more dense mantle there. This has

244

been common wisdom for at least two decades, as you can see from this oft-cited paper†, and was assumed before then. This made the Earth into a sort of gyroscope or top, and explained its angular momentum. It also explained things like precession.**

But obviously you can't have it both ways. If there is more density at the equator, there can't be less gravity there. The only way to push data into line with the field equations is to propose more density at the poles. But all evidence points to more density at the equator. The mainstream even admits this. So they cannot explain this model from GOCE at all. Both the polar flattening and the higher equatorial density work against them. This new data conflicts loudly with their old theory and data.

This model they have published is also strange. I was so busy criticizing the math and theory under it, I forgot to study the actual model closely. It wasn't until my fiancée came in and read over my shoulder that I realized how strange this model really is. She saw the yellow hump at the top, and I told her that was a gravitational high. She said, "Then why isn't it lower? Doesn't gravity pull down?" You know what, she is right. It is extremely odd that they reversed the highs and lows in this modeling. Why wouldn't they match the deformation gravity is thought to cause on the Earth, just exaggerating *it*? I'll tell you why: Because if they had done that, it would more obviously beg the questions I have just begged of them. I drove right around that, as you see, but they had hoped to deflect my questions by cleverly reversing the model. Either that, or they hoped to create so much confusion most readers couldn't even form a reasonable question.

I have been answered by some that the centrifugal effect overwhelms the effect of gravity, and that is why we see the Earth in the shape it is. Yes, gravity alone would create the potato shape we see under title, but the centrifugal effect trumps that, and the equator is forced out. Unfortunately, that explanation also fails for a number of reasons: 1) In the tidal equations the mainstream has fudged, the centrifugal effect is *half* the gravitational effect. See my first tidal paper, or any mainstream math. How can it be half there and more here? 2) There is no indication that the tensile strength of the Earth is low enough to allow either gravity or the centrifugal effect to cause that much deformation. 3) The mechanics still wouldn't work, even if we ignored 1 and 2. Even if the centrifugal effect does overwhelm gravity, we still have to assume that gravity resists the centrifugal effect. The two have to be in opposition, or the centrifugal effect would increase without limit. In other words, if gravity is not resisting the centrifugal effect, the centrifugal effect is only resisted by tensile strength. But tensile strength cannot resist a continuous force. Once deformation begins, it cannot be stopped. This is because tensile strength decreases with R, and deformation increases R. Why is this important? Because, by the current theory, *both* models of gravity fail. I have shown why the model flat at the pole fails. So the mainstream backtracks and gives me the potato model above, saying, "OK, gravity *is* stronger at the equator, so we modeled it like that. Gravity pulls in more there, so we get more deformation there, or would, without the centrifugal effect." Hopefully, you see the problem there. It contradicts the GOCE model, which tells us that blue is a low, not a high. It also sets us into a feedback loop, since if gravity pulls the equator lower, R then decreases, which increases gravity again, which pulls harder, which pulls the equator even lower. We are in a feedback loop, which could only end in a black hole.

You will say, "Good lord, you have shown that both gravity models fail. But it has to be one or the other right? Either gravity pulls more on the poles or on the equator." Yes, according to mainstream equations, gravity should pull more on the equator, as I showed. But using my unified field, as below, we don't encounter any of these paradoxes. Deformation isn't caused by gravity variations or by the centrifugal effect. It is caused by charge emission. It isn't gravity and the centrifugal effect that are resisting one another, it is solo gravity and charge. The centrifugal effect is real, and will cause a force, but it is the tertiary cause here, and can be ignored in the first instance.

Yes, in my Unified Field Theory, we have a perfectly logical explanation for these polar variations from GOCE, one that doesn't require any jumps of logic or *ad hoc* theorizing or reverse modeling. It also allows us to keep the data showing more density at the equator. We can keep both the old data and the new; we just have to update and extend the old theory. I have shown that all celestial bodies pull in charge at the poles and emit it at the equator. They do this by simple rules of angular momentum and field potentials (statistics). And this charge is real: it has mass. It is a real bombarding field, and the photons that make it up are not virtual or messenger. They have radius and spin as well as mass. Beyond that, I have shown precisely where and how this charge field fits into Newton's equations, Coulomb's equation, and the Lagrangian. In previous papers I have already said that charge coming in the poles would affect the

Unified Field there, by augmenting the total field; in the same way it would damp the total field at the equator. This is because charge and gravity are in vector opposition. In the Unified Field, charge causes a force opposite to that of solo gravity (I still call the total field gravity, to match current wording). At the foundational level, charge is always repulsive, while solo gravity is a pseudo-attraction. But in this problem, we have charge variations. We have more charge coming in at the poles and more going out at the equator. By simple and straightforward mechanics, this makes the total field bigger at the poles.

If you are lost, stay with me. We start with solo gravity, which is dependent on radius only. It shows no variation. All the major variations we see are due to the charge field. So if you are at the same radius from the center of the Earth, your solo gravity vector is always the same, no matter where on the Earth you are. But if you are at either pole, you have charge photons beating on the top of your head, pushing you down as well. This makes you weigh more. If you are at the equator, you have charge photons coming up from below you, pushing on the bottom of your feet. So you weigh less. That is a very raw explanation, but it gives you the basic mechanics. I think you can see that by Occam's Razor, my explanation is unbeatably simple. But can it answer all the other data as well? Here and in other papers, I have shown that it can.

I will hit a couple of the biggest outstanding questions right now. The GOCE data shows other variations than pole/equator. A hostile reader will point out that I chose the best view for my under-title illustration. What about this view, over Indonesia (the light area is yellow)?

248

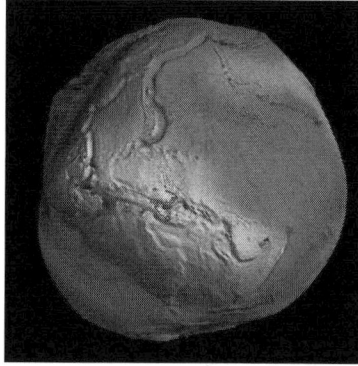

It is clear from analyzing all the views that we have not only the pole/equator variation, we have a land/water variation. And in this particular view, we have even more. We have an extreme high near New Guinea which counters both the land/water variation and the pole/equator variation. What in the world is going on there?

I will hit the first question first. Why the land/water variation? Again, this one is simple. Water, being 2-3 times less dense than land, blocks less of the charge field. More charge coming up means you weigh less. Remember, GOCE is not measuring solo gravity or charge, it is measuring the total effect. It is (unknowingly) measuring the Unified Field. The crust is also much thinner under the water, being as little as 5-10km thick, while the crust under land is five or six times thicker, up to 50km. Again, less blocking from the crust means more charge coming up.

But what about Indonesia? This one is a little more difficult. It stumped me until I noticed that the area in yellow mirrors an area of very high seismic activity. And, if we follow this lead, we find the GOCE maps mirroring the seismic maps all the way round the Earth. The GOCE model would seem to be a stacking of three models: the pole/equator model, the land/water model, and the seismic model. But why would seismic activity tie into this gravity model? Well, our evidence becomes circumstantial here, admittedly, but if we follow our nose, we come to the possible conclusion that something is seriously bottling up charge in these areas. Charge getting through is bluer, remember, so yellow is telling us less charge is coming up. This area of yellow is right on the equator, so it can't be more charge coming down. It must be less charge coming up. If that is so, then we have a new foundational mechanism for Earthquakes. It is possible that the plates beneath these areas are made of denser materials, and that charge is blocked to a greater degree than elsewhere. The charge, being blocked, looks for places to pass, and these places are the seams between the plates, of course. This means that if the sensors on GOCE were even more sensitive, they might pick up bluer or redder lines inside the yellow, that follow the fault lines. For all we know, GOCE may be seeing these already. This model is not finest thing I have ever seen. It looks to be made out of paper-mache. We await more detailed maps or models.

At any rate, I propose that the plates beneath this yellow high are the densest on Earth. We cannot measure deep Earth densities directly, but we may be able to measure these plate densities. The crust is only 5-10km thick under the oceans, so we could compare this plate to the Indian plate (which is the bluest). If I am right, the Indian plate should have the least density, and it should be far less dense than the plates under Indonesia. The Russians have already bored to

250

12km on land, and the Japanese have a ship, Chikyu, that we are told is capable of 7km in the ocean, so this direct measurement is not beyond us.

*http://www.bbc.co.uk/news/science-environment-12911806

**I have shown that this explanation of precession fails, but that doesn't matter here. What matters is that the mainstream *agrees* that there is more density at the equator.

†http://www.sciencemag.org/content/261/5119/315.abstract

Chapter 16

BROWNIAN MOTION

and the Charge Field

Brownian motion is another unexplained phenomenon. You will say that we have long had equations for it, and that Einstein's equations are quite successful. True, but I am not talking about equations. I am talking about a physical explanation. WHY are the particles in motion? We are told that small particles are bumped by molecules, and that molecules are bumped by atoms or ions. But why are the atoms moving in a Brownian manner? It turns out that what we would call balanced atoms and molecules are moving in a random manner, even when they are in what we would call empty space. Even "in vacuum" we see molecules and atoms moving in a Brownian manner. In fact, we have never witnessed particles of any size that were not moving in this manner. WHY?

Once again, without the charge field, there can be no physical answer. We can write equations for the motions, but we cannot explain their genesis. But it turns out that Brownian motion at the lowest levels is more clear evidence for the reality of the charge field. And it turns out that this is one more thing modern physics is hiding from you.

Because the charge field has been hidden in Newton's gravitational equation since around 1687, mainstream physics has missed it all along. They tell us that there is no room in their successful equations for another field, and they dismiss this field as a crackpot's "ether." They tell us that Michelson and Einstein got rid of the ether. But they are wrong. Einstein *extended* Newton's gravitational equation; he did not overwrite it or jettison it. That is why we are still taught it. For this reason, Einstein's own field equations also include the charge field. Einstein is Newton plus time differentials, which means that Einstein's equations are actually unified field equations. The gravitational field has always been unified, which is why Einstein couldn't unify it. He was trying to combine it with the E/M field, but since it already included the E/M field, this was impossible. It is the same problem that string theory has today.

But because modern physicists don't understand that, they continue to try to explain all physical phenomena without a charge field. They can't fit the charge field into their equations (a second time), so they are forced to pretend that the charge field is a sort of virtual field. This is precisely why we are taught that charge is mediated by virtual photons. They can't give the field any real presence, because they can't figure out how to fit the field into their equations. But since I have shown that the charge field is already in the old equations of Newton and Einstein, that problem is solved. We can now give the charge field a real presence. We can also give the photon mass and radius and spin without destroying all the old equations. We don't have to unify anything, since it was already unified.

This being so, we can also admit that the photons we know

about are capable of forces. We already know that from the photoelectric effect and a lot of other experiments, but modern physicists like to teach their students that photons are real sometimes and not real at other times. So we are taught about the photoelectric effect, and are told that real photons are knocking things about. Then the next week we are taught that all quantum numbers and equations are mathematical only, so that quanta don't have real spin or size or position. And the next week we are taught that charge is mediated by magic, either by little plus signs or by virtual photons that can "tell" electrons to move away or move closer. No one ever seems to see this inconsistency of all this. Modern physicists are right about the photoelectric effect, but they are wrong about the others. The photoelectric effect requires that the photon has position, mass, and size. The charge field requires a mediating particle that also has position, mass, and size. And once we give the charge photon mass and size, it is also capable of creating Brownian motion, from the primary level.

All these tiny particles are knocked about by photons. Everything is awash in a sea of photons, and the photons have real momenta. Because they are coming from all directions, they create random motion. And because they are so tiny, they only create a small displacement with each individual hit.

As with my explanation of heat, you will say that I have only taken the explanation back a step. But once again, that isn't true. It isn't true because photons don't experience Brownian motion themselves. Photons (not in a laser) may have a random trajectory, but they do not wiggle about or zigzag randomly. And light can be lasered, which means it can be made to travel a straight line of choice, without wiggling. So I have not regressed in an infinite line of causes, I have found a first cause. The Brownian motion of dust mites is caused by the Brownian motion of molecules; the Brownian motion of molecules is caused by the Brownian motion of atoms or ions, but the Brownian motion of atoms is NOT caused by the Brownian motion of photons. The Brownian motion of atoms is caused by simply by collision with photons.

We can see that photons are not just another step in an infinite regression by looking at velocity. All these larger particles can be slowed down, which is why they experience Brownian motion. Because they do not have a high linear velocity, and because they have a considerable mass, they can move in a Brownian manner. But light cannot be slowed, it can only be deflected. And because the photon is so small, it can move long distances without deflection, even in air. Light can therefore avoid random motion most of the time. Only in certain (and odd) circumstances would light experience something like Brownian motion, and then only if we could slow down time and shrink space a lot. If we pass light through a dense substance, and slow down time, we might get something like Brownian motion. But even so, it is not this sort of motion of light that causes atoms to vibrate. Atoms move randomly due to a bumping by photons, and it is not necessary that the photons be moving randomly themselves.

If this theory of Brownian motion is true, it would imply that Brownian motion could be diminished somewhat in certain circumstances. If we could cohere the charge field present, the atoms would be hit from only one direction, and their vibration would be minimized. On the Earth, this would be difficult to achieve, since the charge field is coming from all directions. It is coming up from the Earth, but it is also coming from the Sun and the Moon and all the planets and the galactic core and so on. However, we can imagine some points in the universe where the charge field would be more uni-directional. In that case, Brownian motion would likely be less. We may propose that nearer the Sun this is true, since there the charge field of the Sun would overpower all other incoming fields. Likewise, Brownian motion should be maximized when the charge field is most random in direction. Brownian motion on the Earth is not maximized, because the main influence is from the Earth. Other influences are present and heavy, but they are not equal to the Earth's influence. But we can imagine some region of space where atoms were surrounded by more equal fields: the Brownian motion must be greater there. The Brownian motion near the center of galaxies must be very great, for instance, even if there is no black hole or superstar there. Any free atom would be surrounded on all sides by large amounts of charge, and would be knocked about quite vigorously. In fact, this is another phenomenon that modern physics has failed to include in its models, and we will look at it again in the near future, to see how it might affect matter in such a situation.

Chapter 17

SUPERCONDUCTIVITY

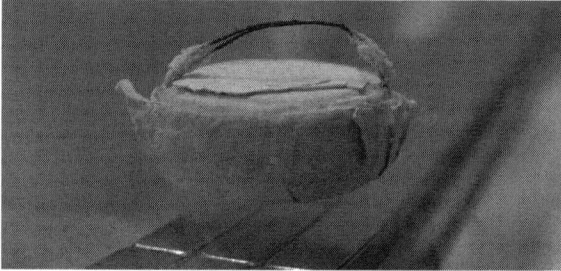

As with so many other things, there is no good theory of superconductivity. Physics now claims to know almost everything, but the number of good physical (mechanical) answers it has to questions is approaching zero.

Superconductivity is currently said to be explained by two theories: the Ginzburg-Landau theory (1950) and the Bardeen-Cooper-Schrieffer theory (1957). Notice first of all that one theory is 60 years old and the other is 53 years old. No theoretical progress in 53 years. It gets even worse when you look at the theories. GL theory is not a theory, it is just a lot of math. Even Wikipedia admits that GL is a mathematical model, not a physical theory, and that it "does not purport to explain the microscopic mechanisms giving rise to superconductivity." So we will pass it by without comment. No, I will pass it by with only this comment: this is the same Landau I critiqued in my paper on the Landau pole and asymptotic freedom. He loves to bury problems under bad math. But since we are looking for an explanation, not math, we won't even pause to pull apart his math here.

BCS theory begins with this:

At sufficiently low temperatures, electrons near the Fermi surface become unstable against the formation of Cooper pairs.

What is a Fermi surface?

The **Fermi surface** is an abstract boundary useful for predicting the thermal, electrical, magnetic, and optical properties of metals, semi-metals, and doped semiconductors.

What are Cooper pairs?

Cooper showed that an arbitrarily small attraction between electrons in a metal can cause a paired state of electrons to have a lower energy than the Fermi energy, which implies that the pair is bound. In conventional superconductors, this attraction is due to the electron-phonon interaction.

I hope you can see that we aren't in the presence of a mechanical theory here, either. The Fermi surface is an abstract boundary, which means the theorists just made it up. We have no data confirming a Fermi surface, and we have no mechanical cause of the surface, so it is completely heuristic. The same can be said of Cooper pairs. Cooper proposes an arbitrarily small attraction, but provides no mechanical cause for it. It is a virtual attraction, in other words, a borrowing of attraction from the void. We see the state of the theory from this paragraph:

Although Cooper pairing is a quantum effect, the reason for the pairing can be seen from a simplified classical explanation. An electron in a metal normally behaves as a free particle. The electron is repelled from other electrons due to their negative charge, but it also attracts the positive ions that make up the rigid lattice of the metal. This attraction distorts the ion lattice, moving the ions slightly toward the electron, increasing the positive charge density of the lattice in the vicinity. This positive charge can attract other electrons. At long distances this attraction between electrons due to the displaced ions can overcome the electrons' repulsion due to their negative charge, and cause them to pair up. The rigorous quantum mechanical explanation shows that the effect is due to electron-phonon interactions.

That is not "a simplifed classical explanation," it is transparent sophistry. Here we have negative charge "increasing the positive charge density." So we are being told that negative charge can INCREASE positive charge, which would be energy from nothing. The increased positive charge then attracts other electrons, so we have electrons attracting other electrons by this mechanism. They "pair up." Each sentence is a new miracle. Not one statement in that paragraph follows from the previous statement.

Good lord, how did we ever come to such a pass that physicists can write and read drivel like that? We are told that the quantum mechanical explanation is rigorous, but if you believe that you aren't paying attention. How could the "rigorous" explanation be good when the simplified explanation is preposterous? Just as an example, we are told that the rigorous explanation depends on the phonon. What is a phonon? It is a quasiparticle. What is a quasiparticle?

261

It is one of the few known ways of simplifying the quantum mechanical many-body problem (and as such, it is applicable to any number of other many-body systems). The most well known quasiparticles are the so-called electron holes, which can be thought of as "missing electrons."

As always, the further you go, the worse it gets. It does not get more rigorous, it only gets more ridiculous. A phonon is a way to fill a hole, in other words. It is a thing that fits the hole in your theory, and then you call that thing a particle. That isn't physics! Also remember that there is no such thing as a quantum mechanical many-body problem or solution, since there are no bodies in quantum mechanics, strictly speaking. There are only probability clouds, remember? This means that many-body problem can't even be defined *mechanically* in quantum *mechanics*. As it is, the solution is just a pasting together of mathematic fudges, as you will see if you study it.

But enough of that. If I wanted to be slapped in the face by a wet fish, I would have gone to the clown market. I want a physical answer to the question, "What causes superconductivity?" If the answer were really that difficult, I would understand all the misdirection. But it turns out the answer is fairly simple. All you need is the charge field. To get a charge field, all you do is let the photon that transmits charge be real instead of virtual. You let it have moving mass, radius, and spin. Since charge is real, it cannot be transmitted by virtual particles with no size or energy in the field. We don't have to propose a phonon to fill a hole in our theory. No, we just have to propose that the particles that our equations give us are real. I mean these old equations:

$e = 1.602 \times 10^{-19}$ C

$1C = 2 \times 10^{-7}$ kg/s (see definition of Ampere)

$e = \mathbf{3.204 \times 10^{-26}}$ kg/s

That means that charge has that mass. The fundamental charge is that much mass per second, which I simply apply to the charge field and the photons that are in it. Charge is then the motion of these real photons, not some mystical attraction or repulsion of ions.

This solves the superconductivity problem because conductivity is defined as the ability of a substance to let charge pass. Obviously, charge will pass most easily when it is blocked the least, and it is blocked the least when particles aren't getting in the way. In other words, charge photons will pass through still matter more easily than they will pass through vibrating matter. A lack of conductivity is explained by photons colliding with matter, and energetic matter will collide with more photons.

We must also remember that in normal circumstances, the field of charge photons is recycled by all matter. It is recycled via spin. Each particle is spinning, and this spin pulls in photons at the poles and spits them out at the equator. But when heat approaches absolute zero, motions slow down near a stop. When motions slow down, collisions decrease, and when collisions decrease, the spins cannot be maintained. The baryons and electrons slow their spins, and nearly stop recycling the charge field. Since the photons are not being sucked in, they are free to pass. The vortices around all particles are diminished, and the field has less resistance. The substance minimizes its collisions, and the charge field therefore maximizes its efficiency. If the charge field is carrying ions of its own, these ions will pass through the substance with minimal collision.

Notice that all my terms here are mechanical terms. My resistance is mechanical, not heuristic, since it is explained directly by real particles and their collisions. The same would be true of potential. Potential in my theory is not an abstract field principle, it is a direct outcome of particle densities. For instance, I will be asked why photons would be sucked into a baryon at the pole? Isn't that a heuristic field statement? No, it is caused by real field densities. Any spherical particle spinning about a pole would tend to fling off excess matter at its equator, because that is where the greatest velocities are. So, IF photons were recycled by protons, they would be emitted at the equator more than anywhere else. Since photons are real, the field density beyond the equator would be greater than at the poles: any external photons passing by the equator would be likely to hit an emitted photon, and would be driven off. External photons passing by the poles would not be driven off, since no photons are being emitted there. This simple mechanism creates potentials: a passing photon has a high potential to be driven off near the equator and a low potential to be driven off near the poles. From a distance, this would create the appearance of attraction at the poles. If people are driven off from all houses but yours, they will seem to be attracted to your house.

In this way, superconductivity can be explained with poolball mechanics. Current theory just doesn't have the right balls.

Chapter 19

What Causes
EARTHQUAKE LIGHTS?

The recent earthquakes in Japan have reminded us of this phenomenon, although it has long been known. It was dismissed by mainstream physics as a myth until the earthquake swarm in the 1960's in Japan, when many pictures were taken. Take note, for this reminds us how new our science really is, and how "skeptical" it is. Skepticism is now sold as a positive attribute of science, but we see again and again that a pig-headed refusal to look at evidence is often sold as skepticism. We will see that again here, where I will show that mainstream physics is refusing to look closely at evidence, or to consider the most likely causes of that evidence.

In a series of recent papers, I have proved beyond any doubt that mainstream physics has buried its head in regards to the charge field, as far back as the time of Laplace and Lagrange. The so-called Great Inequality of Jupiter and Saturn was clear evidence of the charge field, as was the development of the Lagrangian. Both required the presence of a second fundamental field, present at all size levels, quantum and celestial, but physics found a way to ignore this. It diverted itself, through Byzantine channels of argumentation, into ungrounded mathematical models, and it has done this ever since. More recently, we see glaring evidence from the Moon ignored nightly, as the obliterated near crust of the Moon screams of the presence of the charge field, but no one hears. NASA has long known of the negative tide at the front of the Moon, the schematic is published all over the place, but no on sees what this must mean.

Here, we find evidence just as loud being ignored in full daylight. We have lights linked to earthquakes, and since light is an electromagnetic phenomenon, we must have strong electromagnetic waves present. And yet no one appears to realize that this means that seismic waves cannot be, at bottom, simply material waves. We are taught that seismic waves can be body waves or surface waves, each either transverse or longitudinal, but we are never told what causes these waves. We are left with the impression that they are like sound waves, created simply by compression, and that therefore there needs to be no underlying cause of the sort I am talking about. Motion is the only cause, I will be told. But do we have any evidence of that? No. In fact, we have much evidence to the contrary, and earthquake lights are just a small part of that evidence. I will show that seismic waves are not caused by simple motion in the crust, or below it. They are caused by electromagnetic waves.

As with most other science, the science of earthquakes tends to get bogged down in a description of effects, never addressing fundamental causes. In this way, we are taught all about P and S waves, Love waves, Rayleigh waves, and so on, but we are never taught the genesis of all these motions. We aren't taught it, because it isn't known. As with the cause of earthquake lights, it isn't known. At least this is admitted with earthquake lights, where even Wikipedia admits that the various theories are just wild stabs, with no data one way or the other. For instance, one leading theory tells us that earthquake lights might be caused by quartz being compressed, which creates a piezoelectric effect. But no one has shown or even tried to show that earthquake lights are present when crystals are present in abundance, and absent when crystals are absent. That would be fairly easy to do. I assume the data is negative, and that they know that, which is why they avoid looking at it.

As I have already implied, earthquake lights are very conspicuous raw data themselves, which tell us that seismic waves are electromagnetic waves. Yes, these electromagnetic waves then cause material waves in solids and liquids, but the solid and liquid waves are secondary. They are not uncaused, or caused only by motion. Both the motion and the solid and liquid waves are caused by electromagnetic waves. So again, current science is upside down. It tries to explain electromagnetic waves in the earth or the atmosphere by solid waves, when the reverse is true: it is the electromagnetic waves that cause the waves of ground, water, and air.

I will be asked why the Earth should be filled with these electromagnetic waves, and I can only answer "for the same reason everything else is." The universe is filled with electromagnetic waves, and collections of matter like the Solar System are especially rich with them. The Sun recycles the charge field coming from the galactic core, and the Earth recycles the charge from the Sun. All are engines of the charge field, differing only in size and scope. The Earth pulls in charge at the poles and re-emits it everywhere else. So charge is moving up through the Earth all the time. It only requires variation in density to create variations in charge, which then create charge waves. It is these charge waves that drive all other motions in the Earth's crust, its waters, and its atmosphere. Yes, what we call gravity is a player here as well, since this is a unified field effect. But since we already know about gravity, I only need to talk here about charge. Gravity hasn't been left as a hole I need to fill.

But back to the earthquake lights. Given the lights, the most logical theory would be that the seismic waves create the lights directly. That is, the seismic waves do not stop at the level of the earth, they continue up into the atmosphere. Unless we are shown evidence to the contrary, that is the natural supposition, since there is no hard boundary at the surface of the Earth. Yes, we go from solid to gas, but we do not go from solid to vacuum. Even were the seismic wave strictly a compression wave, it would still continue up into the atmosphere, since the atmosphere is material. But if we suppose that the seismic wave is fundamentally an electromagnetic wave, this doubles our bet. The atmosphere is full of ions, even in the lower levels, so the ladder this wave climbs is even clearer and stronger. The earthquake light is then just the seismic wave showing itself in the atmosphere.

I will be asked why we see it sometimes and not at other times. Simply because atmospheric conditions vary. If there are a lot of ions in the air, the seismic waves will light them up. If not, not. To start with, we may assume that earthquakes don't cause or require certain atmospheric conditions: they can erupt under any sort of sky. The seismic waves then conjoin with the given atmosphere, either creating visible light or not. You will say that is just a hypothesis, as bald as any other, but it is confirmed by recent experiments. In 2005 and 2006, several Japanese authors published papers under the titles, "Conditions of atmospheric electricity variation during seismic wave propagation"[1] and "Generation Mechanism of Earth Potential Difference Signal during Seismic Wave Propagation and its Observation Condition".[2] These papers shows just what I am saying: the seismic waves are the direct cause of the atmospheric electricity variations, without any piezoelectric additions, without crystals, without consideration of the ionsphere, and so on. The only reason we don't see the same atmospheric phenomena at all sites is that the atmosphere was not the same to start with. The variation in the atmosphere is dependent on two factors, not one. It is dependent on the local atmospheric conditions, and it is dependent on the quake conditions. The two together create the effects we see.

But the main reason this Japanese paper is confirmation of my theory is that if seismic waves were NOT electromagnetic, we would expect to see NO atmospheric electricity variation during seismic activity. This is precisely why mainstream physics resisted earthquake lights until 1965. With earthquakes as compression waves, there was no good way to explain earthquake lights. They must be a myth. However, please notice that mainstream physics has not been corrected by this data. They have accepted the data but have refused to incorporate it. That they have refused to incorporate it is obvious in the types of theories put forward for earthquake lights since they have been admitted. Not

269

one of the theories since 1965 even tries to link the seismic waves to the lights. All the new theories are the attempt to keep the original theory, simply pasting an addendum to it. First, they assume that the seismic waves cannot be electromagnetic themselves. This being so, the earthquake lights must be caused by some chain of events, but not directly. That is, the seismic waves cause compression in crystals, which cause piezoelectric effects, which then percolate up into the atmosphere. Or, the seismic waves create a disruption in the magnetic field above the quake (by a mechanics not given), and this magnetic field disruption causes the lights. Never is it proposed that the seismic wave simply continues on up into the atmosphere, causing the lights by direct means. That can't be proposed, because that would require that the seismic wave be electromagnetic to start with.

Notice that this chain of reasoning of the mainstream begins with an assumption that seismic waves cannot be electromagnetic waves. Why would they assume that? Well, they assume it because up until the last fifty years, and even into the present decade, no one has wanted to admit the presence of the charge field as a major player except in quantum interactions. If you start to admit the charge field as a major player in quakes, you have to start fielding questions about the charge field affecting celestial equations, and that is the last forbidden topic. The big boys don't want you messing with their celestial equations, which they got straight from Newton, Laplace, Lagrange, and the other demigods of physics.

Unfortunately, we now have recent experiments that show that quakes are indeed caused by E/M potential differences.

One of these same Japanese authors, N. Takeuchi, showed in 1995[3] and 1998[4] that seismic activity was directly linked to electromagnetic potential. In the abstract of the first paper, it says

For all earthquakes with seismic intensities of more than 1 at Sendai we have observed clear variations in the potential difference signal. The observed electric P and S wave arrival times agree exactly with seismic data obtained by the Japan Meteorological Agency (JMA).

And the abstract of the second paper ends with,

The estimated ratio between the earth electrical field (potential difference) and the pressure difference in the ground, i.e., the coupling coefficient, was found to be reasonable based on the streaming potential model.

That's pretty clear, I would say. If you need more evidence of the E/M nature of seismic waves, you will find it just about anywhere you look for it. All you have to do is remove the blinders you have been fitted with by the mainstream, and the charge field becomes a ubiquitous and obvious player in a thousand phenomena and experiments. Just as another (somewhat lighter) example, I will close by reminding you of the well known phenomenon of animals knowing when an earthquake is coming. This has been passed off as a better sense of hearing or a finer sensitivity to tremors, but it is much more likely that animals are sensing E/M fluctuations. We already know that animals use the E/M field in a multitude of ways, and there is even a scientific term for it, "magnetoception." It has been shown in everything from bacteria and fungi to crocodiles and bees, though perhaps its most famous user is the homing pigeon. Recently, in the proceedings of the National Academy of

Sciences, it was reported that animals like cows and deer feed by aligning themselves to the ambient field, and that power lines could disrupt this alignment.* Even humans have shown such sensitivity, though it is weak compared to other animals. Given that, we have a clear and easy mechanism for earthquake sensitivity. It also suggests a new and better mechanism for earthquake early warning. Currently, this warning is done by monitoring P waves or other seismic activity. But I suggest it would be better to monitor the E/M field directly, as animals do. If our machines were very sensitive magnetoceptors, instead of crude motion detectors, we could warn before the main motions started, instead of after they have started. Remember, the E/M field moves much quicker than the matter field, since it is mediated by photons. It is always preferable in such a situation to monitor photons (or electrons) than it is to monitor molecules. Current early warning systems only give us a few seconds of warning, which is fairly pathetic. Just yesterday (April 6, 2011), top seismologists in California recommended spending $80 million on a warning system for the state that would give only about a ten second warning. We have been told in recent stories that the Japanese system gives up to 30 seconds or a minute, but that is pushing the numbers. The minute is the time it took the quake to reach Tokyo from the point of monitoring, not the time it took between local detection and local quake. I am not an expert on this subject, but from doing a bit of research, it appears California could save a great deal of money by installing dogs or crocodiles or pigeons over prominent faultlines, and monitoring their reactions, instead of monitoring P waves. Even better, of course, would be to spend the money developing magnetoceptors or charge receptors that don't reside in the heads of animals.

Many readers will pass by this as a joke, but they just haven't done their research. If you go to this widely available National Geographic article** (which comes up first on a search), you will learn enough to open your eyes, perhaps. I will circle two things in it. One,

There have also been examples where authorities have forecast successfully a major earthquake, based in part on the observation of the strange antics of animals. For example, in 1975 Chinese officials ordered the evacuation of Haicheng, a city with one million people, just days before a 7.3-magnitude quake. Only a small portion of the population was hurt or killed. If the city had not been evacuated, it is estimated that the number of fatalities and injuries could have exceeded 150,000.

Notice that the warning was given days in advance, not seconds in advance. Two,

American seismologists, on the other hand, are skeptical. Even though there have been documented cases of strange animal behavior prior to earthquakes, the United States Geological Survey, a government agency that provides scientific information about the Earth, says a reproducible connection between a specific behavior and the occurrence of a quake has never been made.

See, pig-headedness sold again as skepticism. Could it be that a reproducible connection hasn't been made because US scientists have made little or no effort to make it? Yes, in fact, the USGS admits it, in the article. They made "a few studies" in the 1970's", and then quit. That about sums up the gumption of American science, I would say.

[1] http://onlinelibrary.wiley.com/doi/10.1002/eej.20230/abstract

[2] http://adsabs.harvard.edu/abs/2005IJTFM.125..614O

[3] http://www.sciencedirect.com/science?
_ob=ArticleURL&_udi=B6V6S-3SWKCJ7-
D&_user=10&_coverDate=04%2F30%2F1997&_rdoc=1&_fmt=high&
_orig=gateway&_origin=gateway&_sort=d&_docanchor=&view=c&_s
earchStrId=1710115135&_rerunOrigin=google&_acct=C000050221&_
version=1&_urlVersion=0&_userid=10&md5=b09bf053aa4b4b5a5c64
803b0c386edf&searchtype=a

[4] http://onlinelibrary.wiley.com/doi/10.1002/%28SICI%291520-
6416%28199812%29125:4%3C52::AID-EEJ7%3E3.0.CO;2-P/abstract

[5] http://sciencelinks.jp/j-
east/article/200101/000020010100A0689867.php

*http://www.pnas.org/content/early/2009/03/18/0811194106

**http://news.nationalgeographic.com/news/2003/11/1111_031111_eart
hquakeanimals.html

Chapter 20

BREMSSTRAHLUNG RADIATION

a better mechanism

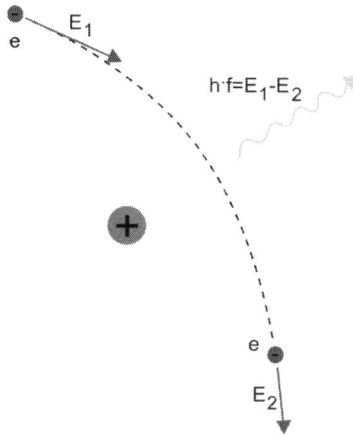

This is a strange one. If you do a websearch on Bremsstrahlung [German for "braking radiation"], you get a large amount of misdirection. Wikipedia, for example, tells you nothing of the history of it, who discovered it, who made up the term, and so on. You also find nothing of the mechanism of Bremsstrahlung. As usual, you are given only the results and some math made up to match the results. But no history and no theory. We have come to expect no theory, but we usually see the heroes of the mainstream mentioned, in a history-bite at least. Here, nothing.

From web sources, you only find that Bremsstrahlung is photons emitted by electrons when they are slowed by near contact with atoms or free protons. But no one ever bothers to tell you how electrons can emit photons. Are electrons little lightbulbs? Are they glass balls full of photons? When an electron emits a big photon, like an X-ray, is the electron diminished by that amount? Don't ask the standard model. Apparently they don't know or don't care.

Same problem if you ask how electrons can be slowed. If they are point particles or, worse, probabilities, how do they feel a slowing force? We are told they don't collide, they only suffer near passes. But if that is so, how is the slowing force transmitted? I thought we were done with force at a distance.

This is where the flunkies all scream at me, "electrons aren't little tennisballs, you idiot! Learn the wave function and get back to us when you have a clue." But notice how convenient it is for them that quantum mechanics has no mechanics. Although they claim to be physicists, the fact that QM and QED are not mechanical allows them to dodge all physical questions. According to them, you aren't allowed to have a natural curiosity, and you aren't allowed to ask sensible and logical questions of them and their theories. No, you are supposed to "shut up and calculate." Memorize what you are taught and then parrot it back to your masters, with a bow and a scrape.

As far as I can tell, the history of Bremsstrahlung is suppressed because it is an embarrassment to the mainstream. Nikola Tesla, who was an outcast of the mainstream for decades, and still is in many ways, was the one who discovered it. Although astronomy.fm still cites Tesla as the discoverer, Wikipedia no longer does. I discovered that Wikipedia recently did cite Tesla, but took down the citation, which tends to support my reading of all this. Maybe some of Edison's descendants are policing Wikipedia, but more likely is that the mainstream simply doesn't like to cite non-tenured engineers, especially engineers who knew more about the charge field in 1880 than they know now. And Grote Reber, an even greater outcast, is the one who discovered Bremsstrahlung as the source of cosmic radio emission, although he preferred to call it free-free radiation. Reber is the ultimate outcast because he was an amateur radio astronomer who beat all the professional radio astronomers to the punch. Rather than thank him for pointing them in the right direction, they have preferred to belittle and ignore him, even after he was proven to be right and they were proven to be wrong (about cosmic radio waves). The term free-free is also still used, but Reber is rarely cited. We can only suppose this is because Reber was a proponent of tired light to the end. The professionals look for any reason to deprive outsiders of their due. Reber should be feted as one of the grandfathers of radio astronomy; instead, the only people who mention him now are on the fringe. I am sure there is pressure behind the scenes to remove Reber's page from Wiki altogether, as a postmortem punishment for not bowing down before their graven images.

As for theory, there is none. Neither Tesla nor Reber nor any of those in between were theorists, so the raw "mechanism" is all we get to this day. The electron emits a photon. But we have no clue as to how or why. If the electron doesn't actually bump the proton, how does it know to emit anything? And if the electron is a point particle, with no real spin, how is energy transfered? How does the photon germinate in the electron, and how is it launched? How can it have more velocity than the electron that emitted it? How does the electron know what energy photon to emit?

These questions may seem too esoteric for physics, but using my new theory of spins, we will find that there are possible answers to them. What is more, the answers are fairly simple.

A couple of my readers have seen this problem coming. They have asked me, "If your protons and electrons are emitting photons all the time, recycling the charge field, then how can they emit photons in instances like this? How could this Bremsstrahlung photon stand out? And what about the energies? Your charge photons are not that much smaller than normal photons. In fact, you have said that charge photons are no different than other photons, charge being only a median or average or something. How does that work?"

What we will see is that the mechanism of Bremsstrahlung, though roughly correct, is not completely correct. The electron is not emitting a photon, *it is becoming a photon*. And this new photon stands out because we have machines set up to track it. That is what these experiments are all about. We have learned where to look and how to look for such photons. We have not learned how to look or where to look for charge photons, since no one has gotten around to seeking them. Generally, you don't find something you aren't looking for.

For example, we know that photons are zipping around all the time. Even without calling them charge photons, we know that every lab everywhere on Earth is stiff with ambient E/M radiation: visible light, radio waves, infrared, the full spectrum. When I am asked how photons emitted in any experiments stand out from my charge field, I return the question: how do photons emitted in any experiments stand out from the known spectrum? They stand out because we are looking for photons of a particular sort in a particular place at a particular time, and we make an attempt to mask the area of other photons of that sort. Well, the exact same answer applies to the charge field. We don't see it with our instruments because we aren't looking for it. In fact, my charge field and this ambient E/M field *are the same thing*. The ambient spectrum IS the charge field. Physics has long know of this field, it just hasn't cared to ask or measure what it is doing, to calculate what it is really capable of.

But back to Bremsstrahlung. Ask yourself this, In all the experiments that record Bremsstrahlung radiation, are the electrons being monitored at the end? In other words, are we quite sure that we have as many electrons coming out as we had going in? No, we know we have a lot of electrons coming out, but we haven't cared to monitor how many. It never occurred to anyone to check. But we already know that all the electrons going in aren't "emitting" Bremsstrahlung radiation, otherwise we would have a lot more radiation than we see. Only the electrons that we think are making the closest passes to the atoms or protons are emitting. This is just to say that I could be right and no one would know it. It is best to show an open door before you walk through it.

My theory has a logical and straightforward explanation for Bremsstrahlung, in that instead of electrons emitting photons for reasons unknown and by mechanisms unknown, I now have electrons emitting photons via easily visualizable means. Instead of an electron and a proton in a void or free space, I have an electron and a proton in a sea of charge photons, and both big particles are recycling this charge via spin. In other words, the spinning spheres are taking in the tiny particles at the poles and spitting them out at their equators. This mechanism works on the same basic principle as an exhaust fan, pulling particles from areas of high pressure to areas of low pressure. You could also explain it as a matter of entropy or statistics. All you need are density variations. But I will not repeat all that here. The important thing is that we now have real spin and real density variations to work with, instead of charge as positive and negative signs or as virtual forces. Yes, the photons as well as the electrons and protons are all spinning, with real angular momenta. With this set-up I have explained many things, and it will allow me to propose a simple explanation here as well. In a free-free meeting of electron and proton, we need to propose that the two particles are spinning in opposite directions. It will not work if the electron and proton happen to be spinning in the same direction. But if they have opposite spins, then their charge fields will also have opposite spins. This helps us explain how particles that only suffer a near approach can feel a real force. Their charge fields are extensions of themselves, and they must feel what their charge field feels.

Nor is this a non-mechanical "field" statement, since I can expand on it if you push me. Since the particles are recyling the field around them, any change in that field will change the particle itself. If the electron's immediate charge field is forced to switch spin, via straight collisions of photons, the electron will also. The incoming charge wouldn't be able to maintain the spin on the electron, and the first thing that would happen is that the outer spin of the electron would be "stripped." This is because the spin maintains the charge and the charge maintains the spin, so that if the spins on the photons are stripped, the spin on the electron will be damped and then lost.

This means that the electron is not "braking" in a linear sense. It is *not* slowing down. It is losing energy, yes, but it is losing angular energy by losing its spin. The spin is what is braking. And the closer the electron comes to the proton, the more spins it loses. If it loses enough spins, it is no longer really an electron. By my spin equation, an electron that loses more than two spin levels actually becomes a photon. That is simply what we call a particle with that number of spins. The electron doesn't *contain* the photon, like our glass jar with photons inside. An electron simply IS a photon with extra spins. We call a photon with two extra spins an electron, and a photon with six extra spins is a proton or neutron.

You will, "By this theory, free electrons and protons should be going c. But we know they aren't." No, my theory never says or implies that all free particles must be going c. They travel at a speed determined by their total angular momentum. For reasons beyond the scope of this paper, particles with angular momentum above a given limit can't go c, since the spin speed begins to conflict with the linear speed. The particle is too big, as a matter of spin radius, to dodge the charge field, and collisions with charge photons begin to slow it.

So, in the free-free interaction, the electron is not slowing down, as a linear matter, *it is speeding up*. It is going from somewhere below c to c, and it is doing this by shedding outer spins. It is the shedding of outer spins, and thus the new smaller radius, that allows it to attain the velocity c. In a word, it becomes small enough to dodge a great deal of the charge field. A particle that can dodge the charge field in this manner is by definition a photon, since it is this dodging that allows it to go c. The photon field is interpenetrable to itself to a large degree, and it is precisely this degree of interpenetrability that determines c, you understand.

As I hope you can intuit, the math of Bremsstrahlung will not be affected by my change in theory. Reber put it this way:

The chances of a close encounter with considerable loss of energy is small. Conversely, the chance of a distant encounter with trivial loss of energy is large. Thus, the spectral distribution will have an inverse intensity-frequency relation. Such is the observed case.

Well, this will apply to my particles recyling a charge field as well, since the odds of close encounter will be the same either way. My theory only gives us a way to explain the transmission of the force. Without a real charge field, either Reber or the mainstream must propose a force at a distance (or, worse, a virtual force).

Now, my mechanism here implies that we should find a sort of anti-Bremsstrahlung if we fire free electrons into anti-matter instead of matter. In that case, the electron would gain energy instead of lose it. It might be given a y-spin on top of its a and x spins, in which case it would be a meson. In fact, this may be the mechanism for muon production in the ionosphere. This would also work with a positron and matter, of course. Currently it is assumed that muons are created by cosmic rays, and this may indeed be one source. But I suggest that anti-Bremsstrahlung may be another.

Given all this, I will be asked, "What about inverse Bremsstrahlung? You have what you call anti-Bremsstrahlung here, and in your next paper you have reverse Bremsstrahlung. But we already have *inverse* Bremsstrahlung, as where a plasma is heated by a laser when the electrons absorb photons. I suppose you want to deny that, too?" Well, I don't wish to deny it, but I do wish to fine-tune the mechanism. I have shown that electrons don't emit or absorb photons in this manner. The electron is already recycling a huge number of photons, so it cannot gain appreciable energy by the interaction with one photon, even if it is of high-energy. What we have is the outer spin of the electron being accelerated by the spins of the photons in question, and, in the case of inverse Bremsstrahlung, we get an electron with a higher frequency. But in some cases, this inverse Bremsstrahlung will become anti-Bremsstrahlung, since the electron will gain enough momentum to add an entire new spin on top of its existing spins, becoming a muon. Hopefully you can see that my anti-Bremsstrahlung is just a more energetic case of inverse Bremsstrahlung.

And there is one other thing to point out with inverse Bremsstrahlung. These physicists must be using lasers made of photons rather than anti-photons. Since light here on Earth is spun by our local field, and since our local field is strongly imbalanced toward matter and photons rather than antimatter and anti-photons, this is not really a surprise, but it is worth mentioning under these circumstances. It is worth mentioning because it could be otherwise. If we were doing our physics on the Moon or Venus or Mars, for instance, we could more easily choose whether we wanted our lasers to be photons or anti-photons, since the ambient field would be nearer balanced and would not (necessarily) be making the decision for us. If we chose to bombard our plasma with anti-photonic lasers, we would then see the plasma cooling instead of heating.

And this leads us into the problem of tired light. Although Reber got a lot of things right, he was still getting some important things wrong to the end, as we will see in the next paper. There I will analyze both the tired light theory and the theory of cosmic expansion. As usual, I will show you some things you haven't seen before. And also as usual, I will pick or create a third side. I will show that both the tired light theory and the expansion theory are wrong.

Chapter 21

The
STEFAN-BOLTZMANN LAW
a simplified derivation

The Stefan-Boltzmann Law is an equation that relates the temperature of a black body to its total radiation:

$J = \sigma T^4$

But if you consult a textbook or the internet or a mainstream physicist, you are never told *why* we have the fourth power here, though that is the central mechanical question. It is passed off as non-mechanical, a coincidence, or unexplainable. But it is a simple outcome of the E/M field.

The problem is that all the textbook and historical derivations have tried to derive the equation from the flat surface of a black body. The current equation is written to apply to radiation per unit surface area, and the surface area is flat. In deriving this equation, the current derivation starts by looking at a small flat surface radiating out into a half sphere. Not only does this make the math much more difficult than it needs to be—since we have to integrate using spherical coordinates, going from flat to curved—but it hides the mechanics of the field. As you see, we are emitting from a flat surface into a spherical field. That is upside down. We should let our definitional black body be a sphere to start with, since that is how the E/M field is emitted in the real world, at the quantum level as well as at the cosmic or planetary level.

In several papers, I have already shown that the E/M field must decrease to the fourth power as it is emitted from spherical objects of any size. It must decrease to the square simply because of the surface area equation. As it is emitted from a spherical body, it moves out into larger shells, and those shells are found by the surface area equation—which has an r^2 in the denominator. But the E/M field is decreasing to a second square because it is expanding inside the gravitational field. Every real sphere creates its own gravitational field, so the E/M field is *always* being emitted into the gravitational field. Because it simultaneously decreases for both reasons, it must decrease to the fourth power.

The Stefan-Boltzmann Law is therefore just the inverse of this fundamental law of the E/M field. If the E/M field *decreases* to the fourth power as it is emitted into larger volumes, it must *increase* to the fourth power with larger temperatures. This is because temperature and volume are inverse measurements. An increase in temperature is operationally exactly the same as a decrease in volume, since temperature increases particle velocity, allowing the particle to cover more distance. If you cover more distance in the same time, you have lowered the effective volume.

We can see this from Charles' Gas Law, which shows the

relationship of temperature and volume:

$$V \propto T$$

The Stefan-Boltzmann Law is just a direct outcome of Charles' Law, and Charles' Law is a tautology—a deduction from the definitions of length and velocity.

You will say, "Wait, Charles' Law is a direct proportion, and you are claiming an inverse proportion! What happened?" What happened is that you are misreading the equation. Yes, with a balloon that is free to expand, an increase in temperature will cause an increase in volume. But if you decrease the volume, the temperature will also increase. In that case, Charles' Law could be written like this

$$V \propto 1/T$$

It is the second case that is experimentally analogous to our E/M field, not the first case. I only wrote Charles' Law as a direct proportion because that is how it is normally presented. If I had written it as an inverse proportion to start with, you would not have complained, you would have written me off as a bumbler. This way, I could show you the logic of my argument.

You will then say, "With the E/M field we cannot increase the velocity, since the velocity is already c." True, but with electromagnetic radiation, we increase the energy instead of the velocity. Energy also acts in an inverse way to volume, since energy always has a mass equivalence. As you increase mass, you decrease volume, because mass must take up volume. As I have shown, mass is operationally or mathematically the same as length^3/time^2, so any increase in mass must increase length, just as with velocity. The result in the field is the same either way.

You will say, "But the photon has no mass. Giving the photon more energy cannot increase its mass, since it has none." I have shown that this interpretation of the photon is used by the standard model only to keep their gauge math working, but they have no proof of it. In fact, by the most famous equation of the 20th century, E=mc^2, all energy *must* have a mass equivalence, including the energy of the photon. Even if the photon has no rest mass, according to this equation—which the standard model still accepts—the energy of the photon must act in the equations and in the field as if it does, so that the energy of the photon takes up real space. We can think of this as *moving* mass equivalence, rather than rest mass, since the photon is always moving. We can say that the photon takes up space in the field due to energy or mass, it doesn't really matter to the math or the field. The same result pertains either way: the effective volume of the field is decreased, proving my point.

Because the Stefan-Boltzmann equation has always been derived from a flat surface, we get the strange constant:

$$\sigma = 2\pi^5 k^4/15c^2 h^3 = 5.67 \times 10^{-8} \text{ Js}^{-1}\text{m}^{-2}\text{K}^{-4}$$

But, of course, if we wrote the equation for a spherical black body, the constant would be different. As you can see, part of that constant is just taking us from degrees Kelvin to watts per square meter. Since the size of a degree Kelvin has nothing to do with a watt, we must get some number transform. But a much more useful and correct equation would give us joules or watts per radius, and in this case the constant might even tell us something interesting and fundamental.

There are two ways to do this. One way is to rerun the derivation from the beginning. But the simplest way is to assume the mainstream derivation is correct, and then convert their number from square meters to a radius. To do this, we just ask what radius would give us one square meter in surface area.

$$4\pi r^2 = 1 = m^2$$
$$r = \sqrt{(1/4\pi)} \approx .282m$$

Then we just substitute into the given constant, like this:

$$\sigma = 5.67 \times 10^{-8} \, Js^{-1}(4\pi r^2)^{-1}K^{-4}$$
$$\sigma = (4 \times 10^{-9}/r^2) \, (Js^{-1}m^{-2}K^{-4})$$

Since r is measured in meters, that will give you an answer in watts per square meter, but your surface is now curved, as in the surface area equation. And you have a variable for radius, so you can immediately find the radius of a particle that would emit that amount of radiation at that temperature. All you have to do is let the r variable have no dimensions. In this equation r is just a number. It's dimensions are separated from it, appearing as m^{-2}.

But let us continue unwinding this equation.

$$\sigma = (4 \times 10^{-9}/r^2)Kgm^2s^{-2} \, s^{-1}m^{-2}K^{-4}$$
$$\sigma \approx (1.2/cr^2)(s/m)Kgs^{-3}K^{-4}$$

$$\approx (1.2/cr^2)Kgs^{-2}m^{-1}K^{-4}$$
$$\approx (1.2/cr^2)(m^3s^{-2})s^{-2}m^{-1}K^{-4}$$
$$\approx (1.2/cr^2)m^2s^{-4}K^{-4}$$
$$\approx (1.2a^2/cr^2)K^{-4}$$

What did I just do? First, I noticed that 4 x 10^{-9} is almost 1/c. Then I reduced the other parameters into acceleration instead. But what does that acceleration variable stand for? It must stand for the acceleration of the surface of our radiating black body, which means it stands for the gravitational field of that body. So we could rewrite the last equation as

$$\sigma \approx (1.2g^2/cr^2)K^{-4}$$

If we let $g = 9.8$, then the equation becomes

$$\sigma \approx (.122g^2/cr^2)K^{-4}$$

You ask, "Why let $g = 9.8$? This is any black body we are talking about here, not the earth." Yes, but I have shown that relative to the earth, all spherical bodies have a gravitational acceleration at their surface of 9.8, including the photon. If they didn't, they would not stay the same size relative to the earth. So if we let g be the relative acceleration of the surface of the body, we can use 9.8 for any given body.

If we use the numbers we already have, we can continue to simplify:

$$T^4(K^{-4}) = 2.56 \times 10^7 r^2 J$$

With that equation, we can relate temperature to energy with just the radius of our black body. J will still be found in watts per square meter, since if we re-expand we would still find the same parameters as before. The transform is no longer a constant, it is true, now that it contains r. Nonetheless, it will be much more useful in this form. We can also use this equation:

$$r = 2 \times 10^{-4}T^2/\sqrt{J}$$

As I said, these new equations will be much more useful than the old ones, since real bodies—especially natural ones—tend to be spheres. The new equations are also much more transparent than the old ones.

Whenever you want to discover the mechanics underneath any equation, first ditch all the modern constants. These constants are there to act as misdirection. Look again at the original constant:

$$\sigma = 2\pi^5 k^4/15c^2h^3$$

A constant expressed by four other constants! And k is not the Coulomb constant, it is the Boltzmann constant, which is two constants, the gas constant and the Avogadro constant:

$$k = R/N_A$$

So the Stefan-Boltzmann constant is actually five constants stacked. They really must want to misdirect us, with that many blind alleys. In other papers I have shown that h, Plank's constant, is hiding the photon mass; and even π is false in kinematic situations. That is why I skipped that expression of σ and went right into the dimenions. As you have seen, it was much easier to solve by unlocking the dimensions than it would have been trying to unlock all the constants.

Some will say, "This equation can't be right, because it implies that all objects of the same radius act the same as radiators of E/M emission. But density must come into play."

No, all this equation tells us is that a perfect black body—our definitional object—must have a certain density at a given radius. It must have this density to be a perfect black body. If it had a greater or lesser density than this optimal density, it would not act like our definitional black body. This is more information that was hidden underneath the old equations. Physicists had thought that perfect black body absorption and emission was due to molecular makeup or some other factor, but this new equation implies it has to do mainly with density. A variety of materials may be able to create this density in various ways, but it is the density that mathematically determines the black body.

And we have one final new discovery, unlocked by these equations. We found that, other than the radius, the transform was composed of g^2/c. I have simplified the derivation specifically to make the mechanics transparent, but what mechanics have we seen here? Why g^2/c? That transform gives us both the gravitational field and the E/M field, the two fields that determine this equation. We simply rewrite that term as $(g)(g/c)$. The first term is gravity, obviously, and the second is the E/M field. The E/M field travels c through the gravity field, so we have to relate one to the other. Yes, g^2/c is the simplest of the simple unified field transforms. I found it only because I was looking for it.

Chapter 22

The

STERN-GERLACH

EXPERIMENT

The Stern-Gerlach experiment was performed in 1922 on silver atoms passing through an inhomogeneous magnetic field. This is the experiment that gave us the current ½ spin values for fermions. I will show that the experiment was badly misinterpreted, and still is, compromising all of QM and QED in many ways.

The first thing we find when studying the experiment is that the physicists assumed going in that the electron was a classical dipole. After the experiment, they decided that the electron was *not* a classical dipole, but they only threw out the "classical" part of that, keeping the dipole assumption. Only by keeping the dipole assumption could they come to the conclusions they did. If they had discarded the dipole assumption, then none of the current conclusions of the experiment would hold, as I will show.

A dipole is a spinning object that is charged differently on each end. The Earth is common example, since we have a north pole and a south pole. Macro-objects that are charged are dipoles, so physicists simply assumed that the electron, being charged, was also a dipole. It turns out this is false, and it can be shown very easily, with simple mechanics.

Modern physics has still not provided us or itself with any field mechanics to explain electromagnetism. They have not explained how force is transferred in the field. The current mechanism is the virtual messenger photon, which can tell a quantum to move nearer or move farther away, depending on whether it is "talking" to a proton or an electron. Since particle physics is in such a naïve state, we should not be surprised that it gets itself into bigger jams every decade.

I have shown that charge requires real photons with real mass and real spin to mediate it. Yes, not only are the electrons really spinning, the photons are, too. Charged particles like protons and electrons are carried along and pushed around by photons, by direct bombardment.

Once we create a real mechanical field, many of the problems of QM and QED evaporate. We don't allow virtual fudges or borrowing from the vacuum or any other magic. This explains the current problem in this way: in macro-objects the dipole is created by the motion of electrons through one pole and not the other. This is already known, and requires no revolutionary theory to confirm it or explain it. But this can hardly be the case with an individual electron. An individual electron cannot be a dipole in this way, since we cannot propose that electrons are moving through or to or around the poles of the electron. The standard model buries this question, having no answer for it and therefore no motive for unburying it. But I have proposed that the electron creates the charge field by recycling charge photons through the poles (and emitting them at the equator). Although they are spinning, photons cannot be defined as having charge, since they *create* charge. In the same way, the electron is the cause of dipoles, therefore it cannot be dipole itself. If the electron is dipole itself, we create a *reductio ad absurdum*, since we must then explain what gives the electron a dipole. We have an infinite line of causes, and therefore no cause.

We cannot assign a dipole to the individual electron for yet another reason of logic. The entire E/M field has been explained as a potential difference between positive charge and negative charge. Franklin defined it like that, with pluses and minuses, Faraday and Maxwell confirmed it, and beneath QED that is still the only surviving pseudo-mechanics. The electron is assigned a minus sign. Well, if we make the individual electron a dipole, then the minus sign has just been transferred to one end of the electron, while the other end is positive. This makes the electron as a whole neutral, like we are told the Earth is. It would also mean that only one end of the electron is attracted to the nucleus, while the other end should be repelled. We can't have that. No, physicists should never have logically been allowed to assume the individual electron was dipole, since it contradicts their first postulates and their fundamental field mechanism (such as it is). The E/M *field* may show dipole characteristics, but the electron itself cannot be a dipole.

If the electron is not a dipole, this changes all the expectations of deflection by homogeneous or inhomogeneous fields. For instance, Wikipedia tells us, "If the particle travels in a homogeneous magnetic field, the forces exerted on opposite ends of the dipole cancel each other out and the trajectory of the particle is unaffected." Well, if the electron is not a dipole, then the trajectory of the particle is also unaffected, but not due to any canceling. There are no forces on opposite ends, so there is nothing to cancel. The particle is unaffected simply because the field is homogeneous. If the field is inhomogeneous, then the silver particles are diverted simply because the atoms have not been made coherent. Some of the atoms are upside down, and the field simply sifts the upside-down atoms from the others.

In other words, in choosing silver atoms for this experiment, "a beam of neutral atoms each having an unpaired electron is used." Well, since we have an unpaired electron, we have a given state with a certain chirality. If that electron is spinning, then it is emitting photons that are also spinning, and we have a normal field. But if we turn the atom over, the electron is spinning in the opposite direction relative to the device, and so are the photons. We then have an inverted field. And so we have *two* possible states. We can have atoms either upside-up or upside-down, relative to the measuring device. If the charge field emitted by the silver is upside down, it must react in the opposite way to the device. Only if we made the silver coherent before the experiment, by making sure all the atoms were upside-up, could we avoid this.

You see, the problem is that we are told the classical expectation of the Stern-Gerlach was a broad band from 1 to -1, going through and including zero. But that is assuming the electron is a dipole, and that is not really a "classical" expectation, it is an illogical expectation. The Stern-Gerlach experiment did not show that *classical* E/M theory was wrong, it showed that *all* E/M theory based on the electron dipole was wrong.

Even with quantum theory, the assumption was still illogical, since with a quantized charge, the experiment should have yielded (we are told) three quantum results at 1, 0, and -1. We do not see that, but it isn't because the spin quantum is less than 1. It is because the prediction of a 0 angular momentum is illogical. It is based on a dipole configuration for the individual electron that does not and cannot exist. No result of zero should have ever been expected, since the spin of the electron cannot be zero. That unpaired electron is either emitting an upside-up field of charge photons or it is emitting an upside-down field of charge photons, but it cannot be emitting both.

You will say that if we have silver atoms that are either up or down, then the charge field will sum to zero; but that isn't how it works. If that were the case, then the *only* expectation would be zero. We could not also have results of 1 or -1.

The spin or angular momentum of the charge field can sum to zero, but only in the case that we have a charge field meeting another charge field head on. But these silver atoms are all moving in the same direction through the machine. In other words, the two values for the charge field are moving parallel, with the same positive vector motion. They therefore cannot interfere with one another, or sum to zero. One atom will be moving in one charge field and another atom will be moving in another charge field, but the charge fields don't mix mathematically or physically, so there is no summing to zero in that way.

You see, it is once again the refusal of physicists to analyze the mechanics of the motions and forces that has doomed them. They have been satisfied with a mathematical analysis, or a cursory physical analysis. They will not go in and actually visualize the motions and forces, to find out the real logic of the situation, because they have been forbidden from doing so. The Copenhagen interpretation has doomed particle physics almost from the beginning.

The entire reason we now have spin ½ particles is due to this experiment, and this terrible interpretation of this experiment. Because the expectation of 1, 0, -1 did not happen, the physicists at the time thought that the quantum could not be 1. They thought a quantum of 1 would necessarily cause a value at zero. Therefore, since zero was not observed, the quantum must be less than one. That is so irrational it is difficult to stomach. I have shown why the value of zero was not observed, and could not be observed; but beyond that, the idea of a quantum of less than one is idiotic. You should always be able to assign any number to your quantum that you like, since the quantum is your unit. It is like saying that you are not free to assign the number 1 to the first integer. If you are not able to assign the number 1 to your first integer, your brain is some terrible state of disarray.

If you don't understand what I mean, just consider that a quantum *is* a one. A quantum is defined as a lowest divisor, something that cannot be further divided. If you are going to start dividing a quantum into halves, you might as well not have a quantum. Once you go below 1, you are into sub-quanta.

If spin is quantized, as I agree it is, then the spin quantum is 1. If you discover a lower quantum, then you should re-assign your integer 1 to the lower quantum, and make your old larger quantum a multiple of that lowest quantum. Any theory that has a spin ½ is a sign of theorists who don't understand the definition of quantum, or of 1.

Another way to say all that is that since there should have been no expectation of zero, the quantum expectations were not split three ways. Since they were not split three ways, the quantum was free to be 1 from the beginning.

The only thing the Stern-Gerlach experiment really proved was that the spin was quantized. It should have been quantized at 1 and -1. But it should have been defined as a real spin. Instead the experiment was used as proof that the spin was not real. Why? We are told that it is because if the spin were real, the energy of the electron would imply a spin speed well above c. More specifically, we are told by Wiki that if the radius of the electron were 1.4×10^{-8}m, then the rotational velocity would be 2.3×10^{11}m/s.[1] But those numbers are found with a bunch of false equations. The writers at Wiki may be aware of the status of those equations, since they don't include them. We only get a footnote to a book published in part at Google books. The equations in question are on page 35. Is it a coincidence that pages 27 to 36 are not included?

Notice that these numbers at Wiki imply that charge is determined by the period of rotation, since those numbers can be found this way:

$$\text{charge} = 1/\text{frequency} = c/\lambda = r/v$$

But of course the radius of the electron is nothing like that large. The classical electron radius is about 10^{-15}m, not 10^{-8}m, so it is unclear why Tomonaga would even be talking about the larger number. In my paper on the Bohr magneton, I corrected early equations, and calculated that the electron radius is closer to 2.24×10^{-17}m. Correcting the corrupted equations $v = r\omega$ and $L=mvr$, we get

$$L = m\omega = h/2\pi$$
$$\mu_B = e\omega/2$$

$$\omega = 2(9.283 \times 10^{-24})/1.602 \times 10^{-19}$$
$$\omega_e = 1.16 \times 10^{-4}\text{m/s}$$
$$v^2 = (\omega^4/4r^2) + \omega^2$$
$$v = c$$

Once we correct the equations, the velocity of the spin is not above c, it is precisely c. So there is no reason to claim that the spin of the electron is "a purely quantum mechanical phenomenon" or that is it some sort of "intrinsic angular momentum."

Those are just two euphemisms for "we don't know what is happening mechanically, so we just refuse to talk about it anymore." Tomonaga's book[1] should be titled, "The Story of Virtual Spin, or How a Bunch of False Equations Doomed Us."

Another thing is interesting about the Stern-Gerlach

experiment. It is known that the Stern-Gerlach apparatus works like a polarizer. If we stack fields in sequence, we get mysterious outcomes as in superposition experiments. We are told these outcomes can only be explained with quantum laws, not with normal mechanics. But just as I solved the problem of superposition and detectors in sequence, I can solve the Stern-Gerlach mystery. Both mysteries are solved with spin, in the same way. We simply have to closely monitor possible spins states, as above.

My analysis also gives us a straightforward mechanical explanation of the Rabi oscillation. This oscillation can be explained only in complex and mathematical terms by quantum mechanics, but with my mechanics of spin it can be explained as a flip of the initial spin state. The applied field acts to reverse the entire quantum, which of course reverses its emitted photons as well. That this is what is happening is confirmed once again by Ramsey's use of a similar device to polarize hydrogen molecules to create a maser. This is just the coherence I was talking about above. If we weren't dealing with real spins, then a polarizer could not possibly work. A polarizer can't work on "intrinsic" or non-physical properties. To polarize you have to have a real polarity, and I would think this is obvious. This polarity is not a dipole, it is just a pole: it is a spin about a real axis, with a real radius.

And once we have a real spin we can analyze, we don't need all the ridiculous Alexandrianisms of von Neumann measurement schemes or decoherence. It is the misinterpretation of this Stern-Gerlach experiment in the 20's that has led to 90 years of stacked misinterpretations and fudged corrections. If we go back and make some simple theoretical and mathematical corrections to these early experiments, we can throw out nine decades of very smelly garbage, collecting in ever greater piles. Something has long been rotten in Denmark, and we now should know it is the Copenhagen interpretation. When will the ghost of the dead King (Bohr) quit haunting us with his false tutelage? When will all the poor Yoricks and their bleached and blanched skulls quit oppressing us with their bad postulates and worse equations?

[1]Tomonaga, S.-I. (1997). *The Story of Spin*. University of Chicago Press.

Chapter 23

The
ELECTRON ORBIT
(the greatest hole in Quantum Mechanics)

It is amazing the things we can look away from, when we really need to. The problem with the electron "orbit" is that the electron and proton have opposite charges, we are told. This causes an attraction, as we know. And yet the electron and proton only seem to attract each other up to a point. The electron is not attracted all the way into the proton itself, it is only attracted to the distance of some shell, near to the proton. This is fairly astonishing, or should be, and yet the standard model completely ignores it. It doesn't even find it necessary to tell us why the electron doesn't continue on in to collision. Wikipedia, for example, conspicuously avoids this question under all headings. In the few instances that QM or QED deigns to notice this problem, it tells us, mostly by implication, that the electron maintains its distance due to its orbital velocity.[1] But this is no answer. Why should the electron, attracted to the proton, suddenly develop an orbital velocity? At what distance from the proton does it decide to start going sideways, and for what mechanical reason? We are led to assume, by the fudgy wording and theory of QED, that electrons must always just miss the proton, as if they always just happen to intersect the proton at a tangent, this tangent being the right orbital distance for the shell. But this is mystification in the extreme. Given particles that are rushing around with opposite charges, we would expect a large number of direct collisions. We would expect a fair number of direct collisions even without opposite charges, wouldn't we? If these quantum particles were asteroids instead of electrons and protons, we would expect direct collisions, no matter their sizes. But if we add charge to the mix, we should expect a highly noticeable number of collisions.

Say you have an electron that just happens to be on an intersecting path with a proton. Is this impossible to imagine? No. Is it possible to imagine that electrons are never on an intersecting path with any proton? No. Now we add charge. The electron is not only on an intersecting path, it is attracted very strongly to the proton. Why does it not hit it?

The standard model has no answer. It just pretends it is not a problem, mostly by ignoring it. Texts never address it. It has been buried. To show this, let us do a search on the internet. Here is a "direct" answer from the University of Illinois physics department[2]:

Why doesn't the electron get sucked into the nucleus since the nucleus is positive and the electron is negative?

-Matt (age 16)

That's a really great question! The picture we often have of electrons as small objects circling a nucleus in well defined "orbits" is actually quite wrong. The positions of these electrons at any given time is not well known at all, however we CAN figure out the volume of space where we are likely to find a given electron. For example, the electron in a hydrogen atom likes to occupy a spherical volume surrounding the proton. If you think of the proton as a grain of salt, then the electron is about equally likely to be anywhere inside a ten foot radius sphere surrounding this grain, kind of like a cloud.

The weird thing about that cloud is that its spread in space is related to the spread of possible momenta (or velocities) of the electron. So here's the key point, which we won't pretend to explain here. The more squashed in the cloud gets, the more spread out the range of momenta has to get. That's called Heisenberg's uncertainty principle. Big momenta mean big kinetic energies. So the cloud can lower its potential energy by squishing in closer to the nucleus, but when it squishes in too far its kinetic energy goes up more than its potential energy goes down. So it settles at a happy medium, and that gives the cloud and thus the atom its size.

That's a really terrible answer, even to give to a 16 year old! In fact, it is just gibberish. The problem is, it really does represent the full answer, the one given to graduate students when they insist upon one (which they rarely do). The full answer has more gibberish, but not more content or logic. QM tries to answer the question by making the electron a cloud or probability, but we must imagine that no matter how probabilistic the electron is, it still must have a negative charge. It cannot have a negative charge far away from the nucleus, acting like a particle, then approach the nucleus and begin acting like a cloud with a positive charge. All this talk of momentum and kinetic energy and HUP is just misdirection. No matter how you represent the kinetic energy or momentum of the electron, you cannot create a repulsion. The dispersion of momentum or kinetic energy into a cloud or probability cannot switch the charge or create a repulsion. And the HUP simply has nothing to say about switching charges or creating repulsions from attractions. This "physicist" should be ashamed to be saying such things, especially to the young.

The problem is, the standard model has no better answer than this, and they know it. Generally, when an adult has the temerity to ask this question, they are not treated with condescension and given this embarrassing answer. Instead, they are browbeaten. They are answered in this way (which I have boiled down from a thousand physics forums posts):

Idiot! Electrons are not spitballs. Go back to school and then get back to me, after you have your PhD.

This saves these PhDs from having to be bothered with basic theory. They can tell horrible lies to schoolchildren and growl at everyone else, and feel safe among their walls.

The truth is, using the wave function to represent electrons instead of representing them as discrete balls does nothing to answer this question. It is complete misdirection. I am not saying the wave function is mathematically incorrect. I am saying the wave function does nothing to explain repulsion. The wave function in no way gives the electron a positive charge, or turns off its negative charge. To do that we would have to change or stop the spin of the electron, and the wave function does not do this. In fact, the wave function was meant to represent some unknown motion or motions or amplitudes of the electron: its complex wave. This wave function does not change its character as the electron approaches the proton, it expresses this character. But for the proton to begin repulsing the electron, the electron would have to change its character. It would have to change its charge in some way.

The answer that includes momentum and HUP is especially dishonest, because it wants us to believe that a probability is somehow an exclusionary force of its own. Yes, the answer from Illinois above treats a probability as a force. Once you boil this argument down, it centers on the idea that making the electron a cloud is enough to explain why it doesn't impact the proton. What you do is smear the electron out into a probability, then give the smear an edge. This edge is then given an exclusionary force, like a material bubble. The electron can't impact the proton because the electron is now a big bubble, and the bubble bounces off the proton! Lovely.

The mainstream physicists always deflect these questions by screaming that the electron is not really in an "orbit." It is in a probabilistic cloud. But a probabilistic cloud does not magically become repulsive just by becoming probabilistic. Probability math should not, and does not, switch the charge on the electron. The electron can have all the wave motions and functions and amplitudes and smearing it wants to have, but becoming a wave or a smear does not bypass this fundamental problem. This is because neither waves nor probabilities are automatically repulsive. There is no theoretical reason to believe that either waves or probabilities are physically exclusive. If they were, then protons and neutrons, which according to QED (see De Broglie, Pauli, Gell-Mann, etc.) also have wave and probability characteristics, could never fit into the nucleus. Their bubbles would repulse each other, and the strong force would have to overcome not just E/M, but also probability repulsion.

This is also proved by the photoelectric effect and many other experiments. The photoelectric effect works both ways: if the photon acts like a particle, the electron must, also. Both the photon and the electron must not only have a discrete energy, they must have discrete positions, otherwise the data would not work like it does. Once again the standard model tries to fudge over this fact with probabilities, but a mechanical explanation requires that both the incoming and outgoing particles must have real position at impact. Energy transfer cannot take place mechanically between probabilities, since probabilities can only work mathematically. The photon must physically hit something, and you cannot hit a probability.

Modern physicists pretend that the HUP has something to say about this, but it doesn't. The HUP addresses the math of QM, not the material. The HUP is an operational rule, nothing more. In fact, even if we accept the HUP as applying to the electrons themselves, it fails to deflect all these questions. The HUP states that we cannot measure position and momentum at the same time, achieving accurate values for both. But it also states that if we don't care about momentum, we can measure position VERY accurately. Just look at the equation, $\Delta X \Delta P \geq \hbar/2$. If we make ΔP very large, we can make ΔX very small. In this case, the nasty question raises its head once more: why doesn't a very small ΔX impact the proton or nucleus?

Quantum physicists can make up all the longwinded, jargon-heavy, illogical explanations they want, but there is a very simple explanation that requires no pettifogging, no faith, and no paradoxes.

It used to be that a person who could come up with a simpler, more transparent explanation was a better physicist. That is no longer the case. Now the person that can come up with the more convoluted, mysterious, wordy, and illogical explanation is the better physicist, since such an answer must seem more "profound."

Of course, I would not bring up this problem and treat the opposition with such contempt if I did not have a better answer for it. Fortunately, I do, one that happens to be very simple and direct, as well as mechanical. I have shown in a series of papers that if we make the charge force mechanical, we must get rid of the messenger or virtual photon that is now said to mediate it. We must replace that virtual photon with a real photon, and give it mass equivalence. Moreover, we must make all force repulsive. There is simply no way to explain attraction mechanically, so we give up on attraction, at the foundational level. Underlying both electricity and magnetism, we have the charge field, or what I now call the foundational E/M field. Although electricity may be either positive or negative, the foundational E/M field is always positive. It is always repulsive. This means that all protons and electrons are emitting real photons, and that all protons and electrons are repulsing all other protons and electrons, via simple bombardment. Attraction is explained by noticing that protons repulse electrons much less than they repulse other protons. In this way, the attraction is a relative attraction. Relative to the speed of repulsion of protons with one another, electron appear to move backwards. If protons are defined as the baseline, then electrons are negative to this baseline.

Classically, this can be explained by the size difference alone. Due only to surface area considerations, electrons are able to dodge much of the emission of protons and nuclei, and so they seem to swim upstream.

If you want to think of protons and electrons as smears instead of particles, be my guest: it doesn't change my analysis at all. Larger smears repel smaller smears less than they repel each other. Smears have size just like particles, and electron smears must have smaller or less dense smears than protons. Or, probability smears of electrons must have less flux, or whatever. However you want to define or imagine the electron, the electron must have more space in or around its probability smear, which means my analysis must hold in any possible field. This being true, I think it is much preferable, from a theoretical viewpoint, to talk of discrete particles. Talking of smears adds nothing to the fundamental theory, and, in fact, often throws a blanket over it.

This explains our current problem in a very direct manner, since the orbital distance or shell or level that the electron ultimately reaches is determined by the distance at which the electron is no longer able to dodge the emission of the proton. If we think of the electron and proton as spheres, it makes this very easy to see (we can think of the spheres as probability clouds rather than particles, if we like, but it does not change the mechanics in any important way). The proton is emitting at a constant rate, we assume. But due to spherical considerations, the emission field must dissipate with greater distance from the center. Which is the same as saying that it gets denser the closer you get to the proton. The electron simply continues to fall nearer the proton, until the field density of emitted photons gets great enough to stop it. At this point, a level of equilibrium is reached. The proton has always been repulsing the electron, but now the electron gets close enough that the proton can stop it from coming nearer. At greater distances, the field density of photons was not enough to stop the electron, but now it is great enough. It is that simple.

Think of it this way, if you like. Let us say you live inside a proton, and you have a little window you can look out of. You are very private, so you have an ingenious intruder system. You have guns mounted all around your spherical "house", but instead of firing bullets, they fire basketballs. All your neighbors are protons, and you have found that you can keep these drifting neighbors away using these basketballs. By long experience, you have found that using a given rate of fire, these neighboring protons never get closer to you than 100 feet. You have also found that at 100 feet, these neighboring protons have an apparent size of one foot. At that distance, they can't really tell what you are doing and you can't tell what they are doing, so you are satisfied. Everything is great until an electron moves into the neighborhood. The problem is, he is a lot smaller and he can navigate the gaps between basketballs. He can only move in a straight line, so many times he gets hit and you keep him away. But over time, by trying again and again, he is able to get quite near. After long years of this annoyance, you find from your records that this electron is able to get 10 feet from your house, but no nearer. Here is the question: how big is the electron's apparent size at 10 feet?

You don't have to ask Marilyn; I will tell you. The answer is: one foot. The electron can defy the field until he reaches the point of optical equivalence to the neighboring protons. At this point the pressure of basketballs on him at ten feet is equal to that on the neighboring protons at 100 feet. Or, to say it in another way, if two basketballs per second hit the protons at 100 feet, two basketballs per second will hit the electron at 10 feet.[3]

Now, the point of this story was not to imply that the proton is 100 times bigger than the electron or to imply that the proton is a simple sphere, any more than it was to imply that little people live inside protons. The point of the story is to show that there is a logical variant to the standard model explanation. We do not have to believe in opposite charges causing attractions or repulsions. We do not have to believe in messenger photons that are capable of "telling" quanta whether they should move nearer or farther away. We can propose a simple bombarding field like this and use it to explain protons repelling and also to explain electrons coming close to the protons. One of the great benefits of this new theory (and there are many many others) is that it explains all at once why the electron does not fall into the proton. It does not collide because it was never attracted to the proton or the nucleus in the first place. Its distance of exclusion is simply much less, based on its size.

Also notice that we can throw this theoretical switch without affecting most of the math of QM and QED. Yes, it will require some fundamental changes, but the bulk of the mathematical content of the wave equations is unaffected.

More than anything, this is a shoring up of the foundations, not a critique of the math. Above all, throwing this switch opens up the road to unification, as I have shown elsewhere. Quantum physicists can proudly keep much of their edifice; but now it is possible to unify that edifice with gravity, in a transparent manner, without the need of strings or other esoterica. It is also now possible to answer the simple questions of high school students without telling embarrassing fibs.

[1] http://answers.yahoo.com/question/index?
qid=20060620215656AAW82ZS

[2] http://van.physics.uiuc.edu/qa/listing.php?id=1226

[3] This short answer assumes that the electron and proton weigh the same, and so "feel" the same force or pressure. They don't, so the answer is incomplete. For optical equivalence to work, we would have to include the gravity field here, as well as the foundational E/M field, and I haven't wanted to get into that. Gravity is present at the quantum level, so my answer is strictly correct. Once we include gravity, all we have to do is assume that the proton and electron have the same density. In which case the falling off of gravity exactly offsets the difference in mass. The repulsive force is 100 times less, per unit area; but the "attractive" force of gravity is also 100 times less, so they cancel.

Chapter 24

The

ELECTRON RADIUS

as a Function of c

I discovered something astonishing today. I was re-reading my paper on angular velocity, checking for errors, when I began reconsidering my main equation, and what it might mean beyond what I had already written about it. Remember that I had shown that the old angular velocity equation $v = \omega r$ was wrong. Due to an error of Newton, or an error interpreting Newton, the equation had gotten skewed. Instead of $v = \omega r$, the equation should be $v = \omega/r$.* This allows us to get rid of the moment of inertia, which I showed is just a fudge factor. It also corrects the Bohr radius and a thousand other things. But what I hadn't yet done is actually insert some numbers into the equation. This is always a big eye-opener.

In a different paper on the Bohr Magneton, I had calculated the radius of the electron by other (but related) means, finding 2.244×10^{-17}m. And in still another paper, I had proposed that c^2 in the famous equation $E = mc^2$ was another scaling constant, taking us from the size of the photon up to a larger field size. I explained that we were scaling a local wave—local at the photon level—up to our own level, where we were measuring it. The scaler was c^2 because the linear motion stretched out the local wave. If the linear speed is c and the spin speed is 1/c, then the difference between them is c^2. That is where the number comes from. But I implied or perhaps even stated in that paper that we are scaling the photon when we do this.

It turns out that is not quite right. It turns out that we are actually scaling the electron up when we do that, and that the spin speed 1/c applies to the electron, not the photon. How do I know that? Watch this:

If we take the equation v = ω/r and insert c for v and 1/c for ω, we get r = 1.11×10^{-17}m. Look familiar? That is almost exactly half my electron radius. We can confirm this method by instead using c for v and $1/c^2$ for ω. In that case, we get r = 3.7×10^{-26}m, which I have shown is about the radius of a photon (photons come in a variety of sizes, depending on many spins they have).

This is important because it tells us more about the equation $E = mc^2$. I had shown where the c^2 came from, but only in part. I had not seen that it could also be assigned to a radius. With this new information, we can rewrite Einstein's equation like this

$$E = mc^2 = m/r_e$$

This tells us that the energy of any given mass is always a function of that mass relative to the electron radius. Or we can rewrite the equation like this

$$m = Er_e$$

That tells us that any mass is some number times the radius of the electron. By that way of looking at it, energy itself becomes a scaler. Energy is not really energy, it is just the number of electron radii involved in the event. Or, we can continue to calculate. Since I have just shown that mass and radius are a function of one another, with the proton mass just the square of its radius, we can rewrite the last equation like this

$$m = Er_e$$
$$m_e = Dr_e^2$$
$$r_e = \sqrt{(m_e/D)}/2$$
$$m = E\sqrt{(m_e/D)}/2$$

The constant D is the Dalton, which is just the number 1821. I use this number as a scaler between the electron and proton, for reasons I explain in that other paper. This last equation allows us to express the mass of any object as a number of electron masses involved.

$$E = 85.336m/\sqrt{m_e}$$

This allows us to calculate the energy of any mass as a multiple of electron masses.

It also allows us to see where the Dalton comes from.

$$D = m_e c^4/4$$

Physicists have never understood where these numbers come from, but the Dalton, also called the atomic mass unit, comes from this equation.

If you go to the standard model, you find that the "classical electron radius" is calculated from this equation:

$$r_e = e^2/4\pi\varepsilon_0 m_e c^2$$

But I have shown how and why that is about 10^2 too large. It turns out the equation is too complex. The radius of the electron is just one over c squared. I have also shown that the permittivity constant is mis-assigned to space. The constant is not an attribute of space, since space has no attributes; it is the gravity field of the proton. Since the equation contains both the gravity field and the charge field, it is another unified field equation. And since charge is dimensionally the same as mass, we can reduce:

$$e^2/\varepsilon_0 m_e = 1$$

You will say that in the current equation, that fraction is about 252. Am I saying that one of these constants is wrong? No, I am saying the current equation is wrong. They have left out a constant, and so they have the wrong number for r. The equation should look like this

$$r_e = e^2/4a\pi\varepsilon_0 m_e c^2$$

where a is the number 252. You see, they forgot to scale the electron to the proton. Since ε_0 applies to the proton, they have both the proton and the electron in the same equation. So they need to scale one to the other with the Dalton, 1821. But if we put that number into the equation, we are still off by 7.22. Well, that number 7.22 comes right out of another paper of mine, since it is the difference between a moving electron and an electron at rest. The electron in this equation is moving relative to the proton, therefore we have to use the moving electron. The full equation is

$$r_e = e^2/(1821/7.22)4\pi\varepsilon_0 m_e c^2$$

And that reduces to $1/c^2$.

What I have just shown is that the current equation is wrong. A field scaling constant has been left out, which makes the equation misfire. Since both the proton and electron are in the equation, we have to scale one to the other. This explains directly why the current equation gets an electron radius that is 252 times too large. The scaling constant is a = $1821/7.222$. Once we correct the equation by adding the scaling constant, the equation reduces to $1/c^2$.

Still, this begs other questions, like why is the spin speed of the electron $1/c$? Is that really what the equation v = ω/r is telling us? It can't be, because the electron can't even go c. So why is the electron radius $1/c^2$?

To answer this, we have to go back to my unification of all quanta. I have shown that the proton and electron and all the mesons are really the same particle, just with a different number of spins. If you take a moving electron and give it one more spin, it becomes a meson; two more spins, it becomes a proton or anti-proton. By the same token, if you take an electron and subtract several spins, you can turn it into a high-energy photon. So the electron is just a common spin level. We could call an electron a very big photon, if we like. Which means that it is almost quibbling when I say the particle we are dealing with is an electron instead of a photon. Rigorously, ALL particle are photons.

The only difference is, at the size of the electron, the particle becomes large enough to begin "eating" smaller particles. The big outer spins are large enough to trap and intake smaller photons, recycling them. These recycled photons then become the charge field.

You will say, "Aren't these photons the charge field both before and after they are recycled?" That's a good question, one I am only now getting close to being able to answer. If larger particles are indeed recycling the charge field, we may ask why. Is it just by accident, as it were, the smaller particles getting trapped only as matter of statistics; or are the larger particles actually feeding off the smaller ones, taking energy from them in some way, and subsisting on them in some way? It is difficult to say. I certainly don't wish to propose that electrons have intention, and it is not even necessary I do so to continue my theory. Even if the trapping of photons by large spins is just an accident, caused by no intention of any electron, the trapping could still function to keep the electron viable. We do not need to say that the electron "lives" on this trapping. We only need to say that the various motions of the electron are maintained by this trapping. Physically, all spins must come from field collisions, and the field collisions of larger particles are simply more complex than those of smaller particles. At a certain level of size, these field collisions create greater vortices, ones that are able to funnel smaller particles through intakes and exhausts. We have such engines at our level of size, and we do not assign intention to them. Gas maintains an engine, or keeps it running, but the engine is not thereby alive. If you wish to assign life to engines or electrons, you won't offend me, but as a matter of physics, it is an external question. Not uninteresting, but external. Physics, as I define it, is mechanics. Such questions are not mechanical.

At any rate, if the larger particles are getting their energy from the smaller ones, then the smaller ones must be losing energy. Which means the photons coming out of the engine must be changed in some way. They must have lost some energy. We may propose that the engine strips the outermost spins of the photon, using it to maintain its own spin. But this means charge emitted is less energetic than charge taken in. Particles aren't emitting a charge field, they are taking the ambient charge field and weakening it in the near vicinity. Therefore, what we call charge is actually a charge LOSS. The charge wind is WEAKER (in some way) near particles than everywhere else. This would create the appearance of an attraction.

Well, you will say, if that is so, can we apply this new attraction to gravity, getting rid of both it and expansion? I don't think so, though it is initially a good proposal. The first reason we can't is that the variance in a primary field can't be larger than the field itself. At the level of the Earth, we know that gravity is much larger than E/M, and that can't be explained if gravity is just a variance in E/M.

It turns out that this energy loss near matter is only a magnetic energy loss. The emitted photons lose spin but not linear velocity, therefore the field is weakened only as a matter of magnetism or spin. The photon density is still higher near matter, and the total linear energy is still higher near matter. Only the spin energy is weakened. Which means we still require a second fundamental motion or field to explain all interactions. We cannot explain everything with E/M any more than we could explain everything with gravity. The unified field must still be dual at all levels, quantum and macro. We require two fundamental fields or forces in opposition to explain the universe.

However, this new finding concerning magnetism may come in useful in later papers. If spins are being stripped to act as fuel for larger quanta, then this may explain other phenomena or data we have not yet addressed. It also may better explain old data, or lack of data. For instance, although I have said that the entire E/M spectrum is probably acting as the charge field, many have complained that my charge field is a poor hypothesis in that we haven't detected it directly. Especially problematic is my calculation that the field should peak in the infrared. I have been told that we don't measure a ubiquitous field at the infrared level. Well, we do, both in heat and in cosmic background radiation (which peaks in about the right place). Black body radiation also confirms my hypothesis. I have shown that many photons assigned to other causes or functions are probably acting as the charge. We had long recognized their existence, we just hadn't found a basic function for them. But if that is not satisfying to some of you, I propose that by this mechanism I have uncovered in this paper charge photons are being partially de-magnetized and de-spun by matter. Since most of our experiments are on the surface of the Earth, this would explain a lack of detection. Our detectors commonly use magnetic fields for detection, and temporary charge loss near matter would explain many things, including energy deficits and inability to "see" the charge field.

In this case, space would have a higher magnetic charge than matter (this charge being the spin of the photons). You will say that if that is so, we would know it, but that is not necessarily so. Since space has little or no matter in it (no ions), there is nothing for this charge field to work upon. Our machines currently detect magnetic fields by detecting the presence of ions. Our machines cannot detect the fundamental E/M field or charge field, except in the presence of ions, since it is the ions they are calibrated to detect. A magnetic field without ions would be undetectable. Which is another way of saying we currently have no way to measure the inherent or photonic magnetism of space. Our machines could hardly be calibrated to detect the spin of small photons, when the standard model does not even know or admit that photons have real spin.

But let us return to the original question. Why does the spin speed tend to be the inverse of the linear speed? I have not answered that in other papers or here. Well, notice that according to the equation $v = \omega/r$, we are not being told that the spin speed of the electron is $1/c$ or that the linear speed is c. We are told that a spin speed of $1/c$ is equivalent to a linear speed of c, given that radius. We aren't being told anything about the actual linear velocity of the particle, we are being told the tangential velocity of a point on the surface of the spin. In other words, the velocity v is not the linear velocity of the electron, it is the linear velocity of a point on the spin tangent. At that radius r, an angular velocity of ω will give us a linear velocity at the tangent of v. We can interpret that to mean that a particle of radius r and angular velocity $1/c$ will cast off or emit a particle from its outer spin at a velocity of c. And we can interpret that to mean that since we see photons travelling at c, and since we propose they are emitted by electrons (as well as protons and other quanta), the electron must be spinning at 1/c. This tells us nothing about the linear velocity of the electron, it only tells us that if the electron is emitting, it is emitting at that radius and that spin velocity.

Fair enough. If that is true, then we should be able to calculate a spin velocity for the proton, by the same equation:

$$v = \omega/r$$
$$cr = \omega$$
$$\omega = 1.23 \times 10^{-5} m/s$$

The angular speed of larger particles is greater than the angular speed of smaller particles, which is precisely why they have more energy.

Still, why is the electron special? Why is it's spin speed 1/c? We can't just accidentally have a fundamental particle with a spin speed of 1/c. No, all our fundamental particles have spin speeds that are simple fractions of c. Just look at the spin speed of the proton I just calculated, $\omega = 1.23 \times 10^{-5} m/s$. That is also not an accident:

$$\omega = 1.23 \times 10^{-5} m/s = 2D/c = m_e c^3/2$$

This takes us back to my spin quantum equation, whereby all spins are multiples of 2, based on gyroscopic rules. Because all larger quanta are built on photons, they must be multiples of the photon. So we can use c and simple equations like this to build any particle, showing both its size and its spin.

But again, why does the electron happen to have a spin speed of 1/c? Because it is one level up from the photon in mass. It is several levels of spin up, but only one level of mass up. What I mean by that is that to find the photon mass, we can divide the electron mass by c, giving us 3×10^{-39}kg. True, in another paper I calculated the photon mass as 92 times larger than that, but photons come in different sizes, as I have said.

Even G is a multiple of c. G = 1/50c

That is not a mathematical coincidence. Simple fractions like 1/50 or 2/100 tell us that our numbers are closely related, and G is simply a function of c, as I have shown elsewhere. Just as D is the scaler between electron and proton, G is the scaler between photon and proton. And, as you have seen here, they can both be written in terms of c.

*For those who think this can't work due to units, you should know that my variables are a bit different than current variables. My v is the tangential velocity. The current v_o is defined as the tangential velocity, but it isn't. It is the orbital velocity, $v = 2\pi r/t$. Since that is a curve, it can't be a linear velocity. My ω is also different. Although it is an angular "velocity", I don't measure it in radians. An angular motion is a curve, therefore it is an acceleration. And so my units do resolve.

Chapter 25

More Proof of the Reality of the

CHARGE FIELD

how *e = 1/c*

We are currently taught that the charge field is virtual—mediated by so-called virtual or messenger photons which have no physical presence in the field or space. Of course we should have known that was false from the start. If we are taught non-physical things in physics, we must be off-track. But I have shown that the charge field must be real. I have shown my readers precisely where it exists in Newton's equations, so it is easy to include it in our math. Because it exists in Newton's gravity field equations, it also exists in Einstein's field equations. Beyond finding it in Newton's fundamental equations, I have found it in the real world, in many real-life problems. Most recently I have shown that "dark matter" is really the charge field.

We are told that dark matter outweighs normal or baryonic matter by 19 to 1, and I have shown how to get that number 19 right out of current and longstanding equations. To be specific, I have shown that we can get the number right out

327

of the current number for the fundamental charge. The fundamental charge, which we are told is either the charge on the electron or proton, is currently $e = 1.602$ x 10^{-19} C. Since $1C = 2$ x 10^{-7} kg/s, $e = 3.204$ x 10^{-26} kg/s. If we divide that by the proton mass, 1.67 x 10^{-27} kg, we get very nearly 19. We get 19.19, to be precise. This means that the proton is emitting a charge every second that outweighs it by 19 times. And that means that the charge field outweighs normal matter by about 19 times. Yes, photons outweigh everything else by 19 times. That is what those simple equations have always been telling us. Unfortunately, the equations have been hiding underneath abstract terms like the Coulomb and the statcoulomb and so on, and modern physicists have forgotten how to do simple math like this. They are so busy filling blackboards with quaternions and Hamiltonians and tensors, and lecturing on black holes and the first split seconds of the Big Bang, they have forgotten how to do highschool algebra. One comes pretty quickly to the conclusion that they are either very poorly educated or they are hiding this stuff on purpose.

At any rate, we can spin this simple math out a bit further. But before I do that, I will answer a couple of questions. One reader pointed out that my math shows mass/second, so we have a time dependence here. That is a bit confusing, since mass shouldn't have a time dependence. This is how I answered him:

Yes, my calculations are time dependent, as you say, but you still must be impressed that the 19 to 1 ratio is derivable from simple classical equations, matching galactic data. That has to be more than a coincidence. I think the reason my math is showing a time dependence is that mass is already time dependent itself. Since I have shown elsewhere that mass is actually a motion, mass is also time dependent. This would mean that nothing is really time independent. Another way to say that is that the current 19 to 1 ratio of "dark matter/energy" to baryonic mass already includes a time variable, without anyone being aware of it. Since the time variable always used is the second, my new equations match the numbers of mainstream equations. The only difference is that my equations include the second explicitly, and theirs include it implicitly. Since the charge field is an emission field, it has to include time. Not that time varies as we move from past to future, but that emission is a thing that happens over time. Emission is a process, not a static fact. That is why my equations include the second. The mainstream equations don't include the second, because they are equations of mass, and it is thought that mass is static when it is not. It may be STABLE, but it is not static. Mass is motion, and all motion includes time, by definition.

To say it a third way, I have shown that mass, gravity, and inertia are three names for the same thing. There is a fundamental motion that underlies all three and explains all three. There are several ways to explain that motion, but by far the simplest way is by using Einstein's equivalence principle. In explaining relativity, Einstein used a visualization, where he put an elevator car in space, hovering in some gravity field. The person inside feels a force from the bottom of the car, but he does not know and cannot say whether that force is caused by gravity pulling him down or an acceleration of the car up. Gravity down and acceleration up are exactly the same thing, both as math and as mechanics. The only difference is a vector reversal. In one case, we draw the vector up; in the other case we draw the

vector down.   This was Einstein's own explanation, and I
am adding nothing to it so far.   But this visualization or
thought problem helps us here, because it allows us to
reverse all the gravity vectors in the universe at once (that
addition is mine, not Einstein's).   If we do that, nothing
changes.   Both the math and the mechanics stay pretty much
the same as before.   But gravity is now explainable as a
motion.   Instead of explaining gravity as a mysterious pull,
caused by nothing, we can now explain gravity as a real
acceleration.   To put it even more simply, everything at every
size level is expanding, including the Earth, the universe,
and the proton.   Everything has an acceleration vector out on
every point on its surface, and that vector out explains
gravity, mass, and inertia, all at once.   You will have to read
my others papers for clarification on this, but we can use it
here to see why mass has a time dependence.   If mass is
explained as a motion out of the surface of the proton, for
instance, then mass now has a time dependence.   This is not
to say that mass will change with time.   It is just to say that if
mass is defined by motion, mass must include the time
variable, since motion includes the time variable.   Motion is
meters/second or meters/second squared or something like
that, so all motion is time dependent.

This just means that all mass has always had an invisible "t"
underneath it.   But because that t was constant, we dropped
it.   The masses of all things stay the same relative to all other
things (roughly), so we can drop the time variable.   It is
there, but we drop it as a convenience.   It is implicit.   In this
way it is like the time variable in geometry.   I have shown
elsewhere that geometry ignores the time variable.   If you
draw a triangle in geometry, you do not keep track of how
long it takes your pencil or pen to draw it.   You just draw it

330

and then take it as given. You take it as existing all at once. In the same way, we have come to think of mass as a thing that exists all at once. We take it as a given. But it is not really that way. Mass is time dependent just like everything else. So when I do these equations and find a time dependence, I am not making a mistake. I am writing down the full variables while everyone else is writing down the geometrical simplification.

OK, next question. We are taught that the fundamental charge applies to both the electron and proton. The proton is positive and the electron is negative. But that can't be right, can it, because according to my theory, the proton should recycle more photons than the electron. Very good. My reader who asked this question saw things clearly. Yes, the fundamental charge can't really apply to the electron in the same way it applies to the proton. The electron *feels* the fundamental charge, but it does not emit it. In the first instance, we can get this straight from the equations above. The charge field outweighs the proton by 19 to 1, but it cannot also outweigh the electron 19 to 1. The electron and proton have different masses, so it must be one or the other. The electron is *in* a charge field that has a strength of e, but the electron is not emitting that charge.

Although the electron is not emitting at strength e, it *is* emitting. The electron is also recycling, but it is recyling photons at a rate determined by its mass. Since the electron is about 1835 times less massive than the proton, it will be recyling 1835 times less charge. The electron is emitting only a fraction of the total charge field, so we will ignore it in most of these simplified theoretical equations. The bulk of the charge field is emitted by the larger particles. We should already know this, because if electrons could recycle a full-strength charge field by themselves, then electrons flying through space would create their own full charge field, even when far away from all baryons. But we know they don't just from looking at the Solar Wind. Electrons between here and the Sun don't act like the reverse of baryons. They are deflected by E/M fields according to rules of their own. This is proof enough of my theory of charge.

Now, let's do some more math and discover some more things. Another reader sent me some equations showing that my number for e above is approximately $1/c^3$. Unfortunately, $1/c^3$ is 3.71×10^{-26}, and e is 3.204×10^{-26}. Close, but it is about a 13% error. Could he still be right? Could we show the cause of the error, as well as the reason why e is related to c? Let's find out.

The first thing to do is look at the dimensions. The speed of light is measured in m/s, and my number for e is kg/s. So we need a transform from meters to kilograms. Looks like a stumper, but I have already shown how to do that in another paper. In my first paper on the unified field, I showed that $1kg = 1m^3/s^2$. The basic idea comes from Maxwell, but I took it a bit further than he did. So,

$$3.204 \times 10^{-26} kg/s = 3.204 \times 10^{-26} m^3/s^3 = 3 \times 10^{-9} \, m/s$$

So e is not almost equal to $1/c^3$, it is almost equal to $1/c$. And we have a 10% error, not a 13% error.

This is of very high interest, because it once again proves not only my theory of charge, but my theory of spins. I have written many papers on tangential velocity and orbital

332

velocity. It was one of the first problems I addressed after my early papers on Relativity. I have shown that since the time of Newton, the two velocities have become conflated. Current physicists think they are the same thing, since they were taught that at the limit, one became the other. But this is not true. Newton went to a limit to find the orbital velocity, but he never said that his new orbital velocity was the same as the tangential velocity. If they were the same, his derivation would have been circular. He takes the tangential velocity as given, then derives the orbital velocity. If they are the same, then he has derived his given, which is circular.

At any rate, I developed an equation to find one velocity from the other, using the radius r, and I later showed that at the size of the photon, a tangential velocity of c was equivalent to an orbital velocity of $1/c$. Well, the reason we are finding a ten percent error above is that we are monitoring the charge field as it is emitted by the proton, and the charge photon is a bit smaller in radius than a standard or average photon. This is the equation I have been using:

$$\omega \approx \sqrt{(2rc)}$$

By that equation, the standard photon would have a radius of about 1.85×10^{-26} m. Just put $1/c$ in there for ω, which I have redefined as the orbital velocity (it is currently defined as the angular velocity, but I have shown that angular and orbital are really the same). But I have shown that the charge photon peaks in the infrared, and infrared photons must be a bit smaller than "normal." If we use the orbital velocity above of 3×10^{-9} m/s, we get a value for r of 1.5×10^{-26} m.

What all this means is that the charge photon, if it has a tangential (spin) velocity of c, must have an orbital velocity of 3 x 10^{-9} m/s, which is just below 1/c. What do I mean by that, precisely? I mean that a point on the surface of the photon has a straight-line velocity of c. That is the tangential velocity of the spin. Using my equation to go from tangential to orbital velocity, that gives us an orbital velocity of 3 x 10^{-9} m/s.

And this means that *e* can be expressed as the orbital velocity of the charge photon. But *why* should *e* be expressing the orbital velocity of the charge photon? Because charge is not charge. Like mass, charge is a motion. Just as mass can be written as the motion of the surface of the particle out from its center, charge can be written as the spin of the photon. But why is the fundamental charge a function of the orbital velocity, not the tangential velocity? Because the tangential velocity is a real velocity, and cannot cause a (continuous) force. But the orbital velocity is an acceleration, and can. I have written many papers showing that any curve is an acceleration, and that includes the orbital velocity, of course. We already knew that, historically, but we forget it in many cases. I have shown the equation to get orbital velocity from tangential velocity, but the orbital velocity is not really a velocity. It is an acceleration. And so it can cause a force or a force field. The charge field is a force field. It causes motions. Which is why *e* can be expressed as a function of this orbital *acceleration*.

Let me hit that one more time, for good measure. Go back to your highschool physics class, and remember how you are taught that circular motion is caused by a centripetal acceleration. According to Newton, circular motion is the combination of some straight-line vector, which we call the tangential velocity and which he called the innate motion of the body. In either case, it is just the motion the body had before we turned on the centripetal acceleration. So we have both the tangential velocity and the centripetal acceleration, to cause the circle. But somehow these two motions are added to get an orbital velocity? How can you add or combine a velocity and an acceleration, and get a simple velocity? Well, you can't. Since an orbit curves, it can't be a simple velocity. You probably had some inkling of this in highschool. I know I did. I didn't understand all that I do now, but I wondered how a centripetal acceleration could be an acceleration. It is the only acceleration I was taught where the velocity doesn't change. The number for the velocity never changes, and yet we have an acceleration? We were told that the angle changed, which was enough to create an acceleration, but that never made much sense, did it?

As it turns out, all we were taught was wrong. The orbital velocity is really an orbital acceleration, and the reason it is an acceleration has nothing to do with a change in velocity. It has to do with the fact that the curve is made up of *three* velocities. You will say, "Three?" Yes, and current theory gets this number right, since, as I just showed, they use a centripetal acceleration and a tangential velocity. The tangential velocity stands for one velocity, and the centripetal acceleration is the other two. That is three. So current theory gets that right, but falls apart after that. It fails to tell you that the orbital velocity is a complex acceleration, with three times variables in it. And this keeps you from understanding all the rest.

So, I have hopefully cleared that up. But we still may ask why or how the orbital acceleration of one photon can be equal to 19 proton masses per second. The answer is in that second. For these quantum particles, a second is a huge amount of time. Remember for starters that a photon can go 300 million meters in that time. That's a lot of energy right there. But also remember that the proton can recycle more than 11 billion photons in one second. So the fundamental charge e is not the orbital acceleration of one photon, it is the orbital acceleration of about 11.5 billion photons.

Chapter 26

The

GRAND UNIFIED THEORY

Many have asked me if my theory is a GUT, or a grand unified theory. So far, I have not sold it as one, and that is because I was offended by the GUT's that came before. They were so wrong-headed and presumptuous, I wanted to have nothing to do with the whole idea. But I have had to field so many questions about this, I have decided to write a short paper about it.

A GUT is currently defined as a theory that would combine the four existing fundamental forces, those being gravity, E/M, strong and weak. A successful GUT would combine not only the current theories, but the current maths. Well, I certainly can't claim to have done that, since I have shown that two of those forces don't even exist. I have ditched the strong force as unnecessary, since we have no data indicating that the E/M field exists in the nucleus. The strong force was postulated to counteract E/M in the nucleus, but a better theory is that it does not exist there. I have shown mechanically and logically why it wouldn't exist there, so all the work done on the strong force has just been busywork. This works out well, since the strong force is the most tenuous of the four. The math and theory underlying the strong force are razor thin, and very little is lost in jettisoning it.

The same can be said for the weak force. Although something is going on with the weak force, and it hasn't been made up from whole cloth like the strong force, it turns out that current theory was right to backslide into electroweak theory. The weak force was initially sold as an independent force of nature, but after the Nobel Prizes were awarded, the theorists admitted that the weak force was probably just a comrade of E/M. They were right in that. The weak force isn't a force at all, it is just a fluctuation in the E/M field seen in certain collisions (beta "decay", kaon decay and so on). It is a variation in the charge field, and of course the charge field *is* E/M. The charge field is photons, and the variation seen in so-called weak interactions is mediated by photons directly.

Furthermore, I can't and don't want to claim that I have unified the various mainstream maths. Rather, I have shown that most of them need to be swept out the door. I have no use for gauge math or tensor math or Hamiltonians or renormalization. All that is just a laser show to keep your eyes off the mess. I have kept a large part of Newton's and Einstein's and even Schrodinger's field equations, but I have shown that important parts of all three have to be excised or whipped into shape. With some major nudges here and there, I have been able to unify gravity and E/M. The biggest nudge was recognizing that G was a scaling constant in Newton's gravity equation, and that E/M had always been included in the classical field equations. This alone allowed me to rewrite all the basic field equations, and to correct literally hundreds of lesser equations in related problems and experiments. Another big nudge was the correction to the angular equations, which had been wrong from the beginning. This allowed for another round of important corrections. Finally, replacing quarks with spins allowed me to clean up the greater messes of QED and QCD, simplifying both. Giving the photon spin allowed me to explain magnetism, and giving the photon a radius and a mass allowed me to solve galactic and cosmic problems in a simple and straightforward way. I recently showed that dark matter is actually just the charge field.

So you can see that the long-sought-after GUT was achieved in a way no one predicted. The four forces were unified by throwing one out, redefining another, and realizing that the other two had been unified from the beginning. In the same way, the maths were unified by throwing most of them out and starting over from scratch. I didn't unify renormalized maths with tensor fields, I went back and reformulated a quantum math that was normal to begin with. I did this by throwing out the point and the point particle. And what allowed me to do that realizing that the calculus was misdefined and misused. I had to correct the calculus, and once I did that I no longer needed to renormalize. Likewise with the tensor field. The same correction to the calculus changed the fundaments of the tensor field, so that no matter whether you wished to use curves or straight lines, you could no longer use mass points. The tensor, like the vector, was redefined as a clear differential or length, and that changed all the postulates and therefore all the solutions. The tensor field also had to be corrected by correcting and extending the time and length transforms, and this clarified the entire field once more.

Strictly, my unification is even broader than any previously attempted or imagined GUT's. I have done things they didn't even realize needed to be done. This is because, in order to unify, I saw that it was necessary to correct many longstanding basic equations that had become dogma. The housecleaning was even wider and deeper than anyone imagined, and all sorts of fortified towers have fallen.

That said, I believe I maintain considerably more humility than those who came before, since I make no claims to a complete or final solution. My unification covers a lot of territory, but it is in no way a Final Theory, a complete theory, or, as Bohr said of his interpretation, "The best we can expect." I have cleaned up a lot of messes, but I have only just begun. Even after a decade of work, what I don't know dwarfs what I do know. Neither this unification nor any other takes us much nearer physical omniscience, and we had best admit it. We will not know all there is to know about physics in ten years or ten thousand. I do not claim to have peeked into the mind of God, gods, or nature, since that is all far beyond me. I only claim to have made some important progress in the understanding of human equations and theory.

Chapter 27

My

UNIFIED FIELD EQUATION

CONFIRMED

Abstract: I show that my unified field equation for force, derived several years ago, and my unified field equation for velocity, derived recently to solve the galactic rotation problem [see chapter 7], resolve. This confirms both solutions and both papers, as well as my papers on π.

In my first paper on the unified field*, I derived a relativistic Unified Field Equation, UFE, in the form

$$F = (GmM/r^2) - (2GmM/rct)$$

I did this with simple postulates, and without any curved field, tensors, or other difficult math. In my more recent chapter [7] on the galactic rotation problem, I developed a similar equation for the velocity of a orbiter:

$$v = \sqrt{[(GM_0/r) - (Gm_r/r)]}$$

Question is, can we derive one from the other? In the

galactic rotation paper, I didn't use my Unified Field Equation, instead deriving the velocity equation from first postulates again. But we can see from the form that they are related, both being differentials of the same sort.

The first thing to do is multiply the first equation by r/m, to achieve

ar = (GM/r) – (2GM/ct)

Then we take the square root

v = √[(GM/r) – (2GM/ct)]

If m_r/r = 2M/ct, we have a match. We can rewrite that equation as

m_r/M = 2r/ct

That implies that the orbiting mass is dependent on the central mass, something current equations deny. Let us solve for t, to see what time that is.

t = 2rM/cm$_r$

If we solve for Jupiter, t = 5,466,035s

Since that isn't the time for one orbit, we have more work to do. It turns out we are off by a factor of 68 in this time period, which should be a factor of 64. How do I know it should be 64? Because in the equation we are comparing the square root of the time to the velocity. The square root of 64 is 8, so the relationship of the velocity to the time is 8. In another paper, I have shown that the tangent is equal to the radius only when we are at 1/8 of the circle. Therefore, the time here is the time for the orbiter to travel 1/8 of the orbit. Only after that time will the tangent equal the radius, and only at that time will the arc equal the tangent. If the arc equals the tangent and the tangent is defined as the velocity, then we can find the velocity straight from the radius. Since these equalities work only at 1/8 of the circle (rather than at a limit, as was previously thought), the time period in this equation must apply to 1/8 of the circle.

You will say we are still off by a factor of 1.06. But as I showed in the chapter [21] on the Stefan-Boltzmann Law, that is the difference between 4 and π, at the fourth root. $4\sqrt{4}/4\sqrt{\pi} = 1.06$. Here, we have 8 as the square root of 64, and the time as the square root of the velocity, so we are at the fourth root. You will notice that π doesn't exist in my equations, although my equations can include an orbit. If we want to insert current orbital numbers into my new equations, we have to transform between π and 4. The number 1.06 is the transform here.

Therefore, the equations do match. As long as we define t as the time for 1/8 of an orbit, we can derive one equation from the other. This verifies both equations, as well as my jettisoning of π in the kinematic circle.

*http://milesmathis.com/uft.html

Chapter 28

The GREAT NEUTRINO MUDDLE

I began getting swamped with emails within hours of the neutrino problem hitting the mainstream press. I didn't take it too seriously, because I intuitively felt that this was a minor problem of math, not the gigantic physics-ending problem it was being sold as. After looking at the paper at ArXiv* for about ten minutes, my intuition was proven correct.

The problem was reported by researchers at the OPERA experiment in the Gran Sasso laboratories, in collaboration with CERN. These researchers reported an average neutrino speed 100.00248% that of c, which is about 7,440m/s *over* c. This corresponds to a margin of error from expectations of $(v - c)/c = 2.48 \times 10^{-5}$, or about 1 part in 40,300. We are told that is outside, or larger than, the margin of error of the experiment (by almost a hundred times!), and alarm bells were set off all over the world.

But it *isn't* outside the margin of error of the experiment. I could see this by paragraph 4 of the introduction (in a paper of about 55 paragraphs, 23 pages, 22 fancy figures and tables, **and 175 co-authors**). The key numbers are here:

The measurement also relies on a high-accuracy geodesy campaign that allowed measuring the 730 km CNGS baseline with a precision of 20 cm.

That is your cause of error, right there, although no one seems to realize it. No one appears to know how velocity, distance, and time relate to one another, so no one knows how to calculate the margin of error here. So I will do it for you. What they have done, no doubt, is divide the first number by the second, to get 3,650,000. The CNGS measurement is correct to one part in 3,650,000, right? The neutrino over-measurement is a lot bigger than that, therefore we have a major problem. But the neutrino measurement is a velocity and the CNGS measurement is a distance. Therefore you can't directly compare the margins of error. Why? Because velocity depends on measuring BOTH time and distance, and you can't measure them both at the same time. As a simple matter of operation, you have to be given a distance from a previous experiment in order to calculate a velocity. We can see that here. We have two measurements, one by CNGS, one by OPERA. One is the distance measurement, one is the velocity measurement. OPERA can't measure the distance and the velocity at the same time, in the same experiment, can they? Think about it.

The velocity is actually a *calculation*. What OPERA is actually *measuring* is the time. They then use the number from CNGS to calculate a velocity. Why is this important? Because it means the margin of error is entering the experiment twice, in two different places. Both the time and the distance have the margin of error, not just the distance.

I will be told that these researchers were using atomic clocks of some sort, with negligible margins of error, but that is to miss the point of how the time of the experiment is actually measured. Even if the clocks have zero margin of error themselves, the clocks are being *used* to measure a distance gap. The distance gap has the margin of error exclusive of any margin of error in the timepieces.

Think of it this way: how do you synchronize perfect clocks that are 730km apart? You have to send something over and back to synchronize them, probably a series of photons. Either that or you just *assume* they are synchronized because they were synchronized when they were together. Either way, your margin of error creeps back in. The assumption is unscientific, and the photon going over obviously brings the margin of error back in.

This means that the margin of error when calculating a velocity in this experiment is not $1/3,650,000 = 2.73 \times 10^{-7}$. It is $1/\sqrt{3,650,000} = 5.23 \times 10^{-4}$, or about one part in 1,910. Since that is larger than 2.48×10^{-5}, we have no problem here. The calculation of the velocity is well within the margin of error.

In conclusion, we see that the 175 co-authors of this paper don't know how to do basic math or physics. This muddle stands as perfect proof of my assertion over the past decade that physics is in serious crisis. It is in crisis not because it has bumped up against insoluble problems, but because it has detached itself from mechanics and from basic math. It has been diverted into mysticism and cloaking math since at least the time of Bohr, and probably since the time Maxwell embraced quaternions, more than 150 years ago. No, we can set that back even further, to the time of Laplace and Lagrange, since I have shown precisely how they ignored mechanics and hid in big equations [see above and my previous book]. I have shown hundreds of instances of top physicists hiding behind esoteric maths, these same physicists unable to do college or even highschool level mechanics and algebra. This is just one more case to add to the already long list.

Also notice how this paper proves my other recent assertion about hiding behind numbers. 175 co-authors? Guys, the only way to really avoid personal scrutiny is to publish anonymously. You might want to run that idea by peer review. I am sure the "peers" will love it, since they already get to review anonymously.

But don't expect my solution to get much traction. The mainstream has been ignoring me on purpose for many years. My readers, after seeing my simple solution, have said, "Oh, these guys will have egg on their faces!" Not likely. Contemporary physicists don't want real solutions, and they want the simple solutions the least. They exist on these big manufactured controversies and the fake temporary jerry-rigged solutions to them. No one is chagrined here, and no one is likely to be. It is all part of the plan to generate continued interest in physics, just like the manufactured dark matter controversy that you see on the cover of every magazine. Notice how well it has worked. Emails are flying all over the world, the internet is buzzing. No one has the time or inclination to read one of my papers, but they have time to waste on this non-problem, with committees working on it for months and months. That is no accident. These committees get paid for working on problems. They don't get paid for solving them.

Which is precisely why I don't bother couching this solution in cuddly language. My goal is not to be accepted by the *status quo*, but to drive around it altogether. I have no interest in being hired as some kind of trouble-shooter for these committees of the inept. When top committees of physicists can't even calculate a margin of error, it is time to close up shop and start over. We have to re-educate a whole new generation of physicists, starting from the ground floor.

That's right. Although my solution proves this miscalculation does not require any revolution in c, physics should not think itself safe from revolution. The revolution has already taken place, although few have heard the news. Like a star exploding in the far reach of the galaxy, the event and the knowledge of the event here on Earth are two different things.

Addendum, 9/26/2011: As I expected, this paper is already being dismissed because "the clocks used were correct to within billionths of a second." Meaning, the mainstream once again can't understand simple logic. If you want to see another glaring example of this, visit my paper on the Galilean transformation, where both the jurors and the editors at AJP couldn't see my point though it was a truism of first-year physics.

Here, what is not understood is that the clocks are not only measuring a time, they are measuring a distance. In measuring the time, yes, they may be nearly absolutely correct, and therefore they don't add to the margin of error of the velocity calculation. But in measuring the distance, they must include the original margin of error once again. To try to make you see this, let me ask you this: "Let us just assume you have the two clocks synchronized, at the beginning and end of the experiment. I won't even ask how you synchronized them. We will let that go for now. But where are the clocks? You have to have them very close to the actual release and capture of the neutrinos, or problems arise, right? You will say, 'Of course, we have the timers tripped by the neutrinos directly,' or something like that. But even so, I can ask, 'OK, how far apart are the clocks?' You either say, 'They are 730km apart,' in which case I have you; or you say, 'It doesn't matter.' If you say it doesn't matter, then I assume you are saying that the clocks could be placed anywhere along the line. They could be hundreds of kilometers away from the release and capture points. But if that is the case, then I can point out that knowledge of the end event must travel from the point of the event to the clock, and that takes time. Photons would have to travel that distance to the clock. Even if you took that into account, you would still have to know how far the photons traveled between end event and end clock, and we would be back to question one. I would have you.

Or, think of it another way. Velocity is distance over time. $V=D/T$. The velocity calculated for the neutrinos requires that we fill in both the numerator and the denominator, correct? We know the distance from CNGS, so that is one experiment, and one margin of error. Total margin of error so far, 2.73×10^{-7}. That margin of error belongs to the CNGS experiment, not to the OPERA experiment. That margin of error now *pre-exists*, even before we start the OPERA experiment. Now we need the time. That is the

OPERA experiment. Second experiment. In measuring the time, we have to know the distance again, since we have to know how far apart our clocks are. Time is just a second measurement of distance, so that we can create a ratio and calculate a velocity. *Each experiment* has a margin of error, due to the distance, and we have two experiments. In the equation V=D/T, *both* the numerator and the denominator have a margin of error. Because a velocity requires two separate measurements, *any* velocity will contain two margins of error. And since we multiply margins of error, not add, the actual margin of error in the OPERA experiment is 5.23×10^{-4}.

Some still won't get it. They will say something like, "What do you mean the clocks are measuring the distance? The clocks are measuring the time and we are given the distance. Clocks don't measure distance, you ninny! Yes we have the margins of error in the numerator and denominator, but the error in the denominator is negligible." Stated with enough surety, that almost sounds convincing, which is why a lot of top physicists can't see the truth here, I guess. So I remind everyone that one clock is not the same as two clocks separated by a distance. Yes, if we had one atomic clock in the OPERA experiment, at no distance from the experiment, the margin of error in the time would be near zero. But you cannot measure neutrinos going 730km with one clock, can you? Those who say this are refusing to study the actual operation of measurement of time. You have two clocks separated by a specific distance, and that distance of separation matters. Not for reasons of relativity, but for reasons of operation. It matters that the clocks are 730km apart, and not any other distance, because if they were at any other distance *they could not measure the time of the experiment.* In this way, the distance between them enters the margin of error in the denominator. The margin of error enters the calculation of the velocity *twice*.

One final way, different from the rest. I will use calculus-talk to try to explain it. Velocity is the rate of change of distance, right? It is the derivative of the distance. If we used primed variables to mark our parameters, the distance is unprimed, the velocity is single primed (and acceleration would be double primed). Well, the only time variables can be compared as regards margins of error is if the variables have the same priming. When we are talking margins of error, you can compare distances to distances and velocities to velocities, but you cannot compare distances to velocities. Why? Because if you compare a velocity to a distance, you are ignoring a level of dependence. Calculus requires dependent variables, remember? They are called functions. Velocity is a function of distance, and a rate of change of distance, and both are telling us the velocity is one step more complex than distance. How? The velocity also has a time dependence that distance does not have. More complexity in a variable automatically tells us it has multiple margins of error. Just as a velocity has two margins of error embedded in it, an acceleration has three. A cubed acceleration has four. Every distance or time constituent in a variable implies another margin of error, since each constituent must be *measured* separately.

Now, admittedly, this last part is not common knowledge, and I shouldn't expect everyone to know it. I was not taught it and I am not aware that it is taught at all. But even without that, scientists should know, as a simple matter of logic, that two experiments imply two margins of error. This late in the history of experiments, things like this should be known. It is incredible they aren't known.

*http://static.arxiv.org/pdf/1109.4897.pdf

chapter 29

PARTICLE PHYSICISTS ADMIT IT:
THE STANDARD MODEL IS DEAD

It's been about a decade since I began announcing the death of mainstream physics. Now, we begin to see the mainstream physicists admitting it themselves, in public and in print. I send you to the *Guardian* newspaper*, London, one of Europe's top newspapers. There we find a physics blog hosted by the *Guardian* called "Life and Physics, Jon Butterworth". Butterworth is a professor at University College London, but more importantly he is a member of the High Energy Physics Group at the Large Hadron Collider. In his post from October 11, 2011, he asks in his title, "Perturbation Theory: **are we covering up new physics?**" The subtitle is: "*A timely award of the J. J. Sakurai Prize acknowledges how hard it can be sometimes to pin down what the Standard Model really thinks.*" Like this subtitle, the body of the post betrays a man with borderline schizophrenia. Notice that Butterworth calls the Sakurai Prize award "timely" at the same time he is questioning the entire direction of physics. This is no accident or poor word choice, since he does precisely the same thing throughout the entire post. He cannot come out and say that the Sakurai Prize and its recipients are frauds, but he suggests it several times, in low and backhanded tones.

He begins by admitting that the central question for him is: Does the Standard Model of particle physics work at Large Hadron Collider energies or not? He can't just come out and say "it does **not**", but he implies it. He tells us,

In general we can't solve the Standard Model exactly. We use approximations.

Then he sidesteps into a discussion of perturbation theory. Unfortunately, this discussion betrays him as well, because it quickly becomes clear, even to a first time reader, that he is hedging. He admits that they don't use approximations, they use computer models and gaming models to push the numbers from experiment into line. But that is the definition not of "solving inexactly" but of "fudging." If the standard model were "approximately correct," that would mean it was correct to within some tiny fraction. But if you study the math and the way the numbers are pushed into line, you find that this isn't the case. Yes, each fudge in a set of perturbation pushes may be small, but the change to the final number is not small. Or, the change to the number may be small, just because they are dealing with very small numbers to start with; but the change in the number relative to the particle they are measuring is huge. It may be thousands of times larger than the particle they are measuring. Again, that isn't an "approximation", it is a gigantic fudge.

Then he admits the biggest pushes come when the strong force is involved. In other words, he is all but admitting that quantum chromodynamics QCD doesn't work at LHC. Readers of my papers would have expected this, because I have shown that the strong force is the weakest part of quantum theory. In my paper on the weak force, I said outright that even electroweak theory was rigorous compared to strong theory, before I went on to show that electroweak theory was wrong top to bottom. And in my papers on strong theory (see my first book), I showed that there is no strong force. I showed how the math was rigged from the first equation to match data.

Butterworth admits,

Aspects of how quarks and gluons are distributed inside the protons we collide can't be calculated from first principles. Neither can the way the quarks and gluons turn in to new hadrons in the end. We have some constraints from our theory, we have basic stuff like the conservation of energy and momentum, and we have a lot of data from other places. But we can't use perturbation theory. The coupling number gets near to one, and 1 x 1 x 1 x ... = 1. This means no matter how many particles you include in your calculation, you don't converge on a solid answer. In the end we have to make educated guesses, or models. And these are always adjustable.

That is pretty clear, I would say. Not only does QCD not work, but it isn't pushable in the "normal" ways. The old method of cheating, perturbation theory, doesn't work. So Butterworth introduces us to Monte Carlo theory, which is just an old random sampling trick from von Neumann in the 1940's. Butterworth wants to glide from there into praise of the Sakurai Prize winners, since they use these tricks to prop up Butterworth's current work, and indeed the work of all particle physicists. In other words, without this new trick, Butterworth and the LHC people have nothing much to go on. They *need* the Sakurai Prize winners, and can't really be seen attacking them. If Butterworth chews up Monte Carlo theory, for instance, he just shoots himself in the foot, because then the LHC experiments are just done in a theoretical vacuum.

So what does Butterworth do? Rather than step up to the plate, he passes the buck. He gives us a link to his friend and colleague at LHC, Lily Asquith. In her *Guardian* blog from almost a year ago, she says,

Again, we are talking here of the Monte Carlo simulations that are provided to us experimentalists so that we can check what we observe against what the theorists "predict". I should digress a moment - what the theorists predict is in fact no longer a prediction. They make a prediction and then they "tune" it, so that it fits our data...

As experimentalists, we are terribly upset about this. Despite the fact that every single one of use has a PhD in particle physics, and thus was, at some point in the not-too-distant past, completely in awe of the fact that these guys can even imagine such a thing as QCD, now that we have *The Best Machine in the World Ever* working and taking data, we are completely disgusted that they have not perfected their understanding of theories that are so ridiculously complex....

Wow. Doesn't give you much confidence in the new cheat, does it?

Butterworth tries half-heartedly to paper over this cheat. Although he admits that "she [Asquith] and commenters worried that we might be adjusting these models in such a way that we actually covered up exciting new physics," he tells us this worry can be addressed. It is addressed by having

calculations of what you know, done with perturbation theory, linked up to models of what you don't know very well. I think of this rather gruesomely as a skeleton of hard predictions inside and squidgy body of best guesses. The body can change shape. You can push in its stomach quite painlessly, but you really know about it if you break a bone.... Anyway, marrying the squidgy models to the rigid perturbation theory is mostly done using Monte Carlo event generators.

And, then, as Asquith put it, *tuning it to fit the data*. Not only that, but reread closely Butterworth's last paragraph. Reading backwards, we find that the "skeleton of hard predictions" is provided by perturbation theory. But didn't he just tell us that perturbation theory didn't work here? How can Monte Carlo methods marry "squidgy models" to something that doesn't work? Remember that Butterworth has admitted in this very post that in QCD

we can't use perturbation theory. The coupling number gets near to one, and 1 x 1 x 1 x ... = 1. This means no matter how many particles you include in your calculation, you don't converge on a solid answer.

Well, Monte Carlo is a random sampling method. If it "doesn't matter how many particles you include," random sampling can't help you. It is also worth pointing out that Monte Carlo methods are all methods of desperation. As Wikipedia puts it,

When Monte Carlo simulations have been applied in space exploration and oil exploration, their predictions of failures, cost overruns and schedule overruns are routinely better than human intuition or alternative "soft" methods.

Hah. Better than intuition, you say? That's hard science for you. Come on! Every mathematician knows that "tools" like Monte Carlo are used only when you've got nothing else to go on and you are flying by the seat of your pants. Wikipedia says they are used "to model phenomena with significant uncertainty in inputs," and a good physical theory shouldn't be so uncertain, should it? When physicists begin using Monte Carlo methods you *know* they are desperate. At Wikipedia, they show how to use Monte Carlo to estimate the value of pi. But we don't use Monte Carlo to estimate pi because we have math that works better. QCD ought to be able to calculate what will happen at LHC, in the same way we calculate pi, with straight equations. The only reason we would use Monte Carlo to estimate pi is if we had equations that didn't work. And this means QCD doesn't work. Why not admit it? Simply by using Monte Carlo, they ARE admitting it, but it would be much cleaner to just say it outloud: <u>QCD is worthless, both as theory and as mechanics.</u>

By not admitting it, Butterworth and all these people ARE covering up exciting new physics: MINE. I have shown how to replace QCD with a mechanical theory and relatively simple math. My equations are not in final or perfect form, but they would require far less tweaking than anything the mainstream has ever come up with. And because they are mechanical, any repairs that are required can be done at the ground floor.

Instead, Butterworth is trying to sell us on the Sakurai Prize winners Bryan Webber, Guido Altarelli and Torbjorn

Sjostrand, who won

for key ideas leading to the detailed confirmation of the Standard Model of particle physics, enabling high energy experiments to extract precise information about Quantum Chromodynamics, electroweak interactions and possible new physics.

Unfortunately, we now know that is false. They have not enabled anyone to "extract precise information" from anything. They have not extracted precise information, they have piled one fudge (Monte Carlo) on top of another fudge (perturbation), and married that to a squidgy.

Butterworth crows that unlike string theory, which "isn't even wrong," the new generators of post-Monte Carlo theory at least "describe data". But that isn't true. Neither string theory nor QCD describe data, since physics doesn't describe data to start with. Physics is supposed to match data or predict data, and the new generators do that only with massive fudges and "tunes". Nothing in mainstream physics, in any theory, resembles in any way the old physics or mechanics, where solid equations describing real particles or interactions were written to create a coherent universe.

Instead, all the top prizes in physics now go to the new sort of equation finessing and computer fudging, including the Nobel Prize. Physicists don't get prizes for doing actual physics anymore, they get prizes for hiding the fact that mainstream physics is dead. Over the past century, physics has devolved from a semi-rigorous discipline into an ever-growing pile of mathematical cheats. Feynman himself tried to warn us of this, telling us that his own cheat, renormalization, was no better than "hocus-pocus". But renormalization is a fairly esoteric and subtle cheat compared to Monte Carlo. By bringing Monte Carlo into QCD and particle physics and the LHC, Butterworth and all the rest are simply signaling behind their backs that they know the jig is up. They may be able to fool the editors at American Physical Society, but they can't fool anyone on the outside. Like Lily Asquith, anyone with a PhD in math or physics—no, anyone with a basic understanding of math or physics—can see the writing on the wall.

*http://www.guardian.co.uk/science/life-and-physics/2011/oct/11/1? newsfeed=true

ABOUT THE AUTHOR

Miles Mathis (b. 1963) attended both Haverford College and the University of Texas, where he graduated *Phi Beta Kappa* and *Summa cum Laude*. Having turned down full scholarships to Rice and UT in engineering, he then turned down full scholarships to graduate schools to pursue his own coursework. He is an autodidact in many subjects, including painting, sculpture, poetry, art history, art criticism, physics and mathematics. He has published over 2,000 pages of physical and mathematical analysis, much of it highly controversial. He claims the distinction of being the only professional artist whose physics books are recommended by a NASA astrophysicist (see the introduction to his first book *The Un-unified Field*).

To read more about the greatest standing errors of physics and mathematics, see milesmathis.com.

To read more about the author,
see mileswmathis.com/bio.html

Printed in Dunstable, United Kingdom